About Island Press

Island Press is the only nonprofit organization in the United States whose principal purpose is the publication of books on environmental issues and natural resource management. We provide solutions-oriented information to professionals, public officials, business and community leaders, and concerned citizens who are shaping responses to environmental problems.

In 2005, Island Press celebrates its twenty-first anniversary as the leading provider of timely and practical books that take a multidisciplinary approach to critical environmental concerns. Our growing list of titles reflects our commitment to bringing the best of an expanding body of literature to the environmental community throughout North America and the world.

Support for Island Press is provided by the Agua Fund, The Geraldine R. Dodge Foundation, Doris Duke Charitable Foundation, Ford Foundation, The George Gund Foundation, The William and Flora Hewlett Foundation, Kendeda Sustainability Fund of the Tides Foundation, The Henry Luce Foundation, The John D. and Catherine T. MacArthur Foundation, The Andrew W. Mellon Foundation, The Curtis and Edith Munson Foundation, The New-Land Foundation, The New York Community Trust, Oak Foundation, The Overbrook Foundation, The David and Lucile Packard Foundation, The Winslow Foundation, and other generous donors.

The opinions expressed in this book are those of the author and do not necessarily reflect the views of these foundations.

Reconstructing Earth

Reconstructing Earth

Technology and Environment
in the Age of Humans

Braden Allenby

ISLANDPRESS

Washington • Covelo • London

The short essays that appear in this book are courtesy of *The Green Business Letter.*

Library of Congress Cataloging-in-Publication data.

Allenby, Braden R.
   Reconstructing Earth : technology and environment in the age of humans / Braden Allenby.
       p. cm.
   Includes bibliographical references and index.
   ISBN 1-59726-014-2 (hardback : alk. paper) -- ISBN 1-59726-015-0 (pbk. : alk. paper)
   1. Environmental sciences. 2. Environmentalism. 3. Environmental management. 4. Sustainable development. 5. Science--Social aspects. I. Title.
   GE105.A39 2005
   333.72--dc22

                                                      2005006241

British Cataloguing-in-Publication data available.

Printed on recycled, acid-free paper

Design by Brighid Willson

Manufactured in the United States of America
10 9 8 7 6 5 4 3 2 1

Julia Trusdale Allenby, 1925–2002

Contents

List of Columns

Chapter 1

Chapter 2

Chapter 3

Chapter 4

Chapter 5

Chapter 6

Chapter 7

Chapter 8

Chapter 9

Acknowledgments

This book, and the underlying conceptual framework of earth systems engineering and management, draw from many different disciplines. Accordingly, I have had the benefit of discussions and input from many people from a number of different communities over a fairly long period of time. I hesitate to try to thank everyone, as embarrassing oversights are certain, but there are a number whose contributions I want to recognize despite that risk.

Perhaps most importantly, I would like to thank Joel Makower, editor of the *Green Business Letter* and creator of GreenBiz, and the National Environmental Education and Training Foundation (NEETF). Many of the essays in this volume were first published in the *GBL*, which is supported by NEETF; without them, this book would not have been. And as long as we're on the subject of institutions, I would also like to thank AT&T, which has always supported my work and treated me well: if all companies were like "T," it would be a better world. If only it were in a less chaotic sector.

A major area of influence is obviously industrial ecology, a field where I have learned from many friends, such as Tom Graedel, Rob Socolow, Bob Laudise, Bob Frosch, Dave Rejeski, Dave Allen, Clint Andrews, Reid Lifset, John Ehrenfeld, Dave Marks, Scott Matthews, Arpad Horvath, Cindy Murphy, Barbara Karn, Alan Hecht, Derry Allen and many others. I have also benefited from many discussions over the years at AT&T with people such as Michele Blazek, Clair Krizov, Joe Roitz, Debbie Petrocy, Hossein Eslambolchi, Larry Seifert, Art Deacon, Alice Borrelli, and Mike Keady. My interactions with the National Academy of Engineering over the years, in particular with Bill Wulf, Bob White, Bruce Guile, and Deanna Richards, have been both stimulating and a salutary grounding in technological realism. I learned much from Max Stackhouse at Princeton Theological Seminary, and Mike Gorman and Matt Mehalik at the University of Virginia, with whom I have had the privilege to co-teach courses in earth systems engineering and management from very different perspectives.

Those who have not written books may not understand why authors always—or always should—recognize their patient families. It's because things such as books, even if they only bear one name, truly reflect joint efforts. Accordingly, I thank Carolyn, my spouse and friend, and my children Richard and Kendra.

Finally, I would also like to thank Todd Baldwin, whose editing of this book was both thorough and intelligent; he did an excellent job, and what was not fixed cannot be imputed to him. I would also like to thank my co-author, Dulce, an orange and white feline whose pawmarks are all over this; if her comments were difficult to decipher, they were always sincere and greatly appreciated, and she is greatly missed.

Needless to say, none of the above can be blamed for the final product, which, as always, is the responsibility of the author. They have done their best over the years to educate me, and if that has proven a bridge too far, it is certainly not for their lack of trying.

Introduction

The Evolution of a Movement

November 1918 marked the end of one of the most horrible human adventures of what would become a century of violence, the First World War. France, which had lost 1.7 million men out of a population of about 40 million, was determined to avoid such a calamity again. She and her allies therefore imposed a punitive peace on Germany through the Treaty of Versailles intended to prevent Germany from ever rearming. On the military front, France built a huge defensive wall, the Maginot Line. This massive fortification stretched along her borders with Germany and Italy, from Switzerland to the Ardennes in the north, and from the Alps to the Mediterranean in the south. Costing some three billion francs in its first phase alone, the line included a vast network of interconnecting tunnels where, beneath the earth, thousands of men slept, trained, watched, and waited—to fight the kind of war that was already obsolete.

For actions have unforeseen consequences, and systems evolve. Thus, the policy behind the Treaty of Versailles encouraged radical German nationalism and a rearmament regime that neither Britain nor France had the political will to stop. The Maginot Line encouraged the German High Command to dramatic innovation, in particular the creation of the coupled tank, air, and troop warfare system known as *blitzkrieg*, "lightning war." The result? When the Germans attacked France in May 1940, it took the offensive only three days to reach the Channel coast at Abbeville; the British were driven from the mainland by June 4; France surrendered on June 25.

It is often the case that an overwhelming victory is followed by a strong desire to stabilize and privilege the new status quo. But such victories tend to discourage continued policy innovation and encourage a reactionary conser-

1

vatism even as change and evolution continue all around. The result is frequently a cultural Maginot Line, as once-powerful and effective policies and ideologies become increasingly dysfunctional and anachronistic in a world for which they were never intended, and in which they no longer resonate. And yet in many cases the underlying purposes for which such policies evolved—say, to prevent further apocalyptic warfare in Europe—remain valid and important. The Maginot Line failed; the European Union has succeeded. The former, reifying the trench warfare of the First World War in massive concrete form, looked to the past; the latter, inventing a new European federalism, looks to the future.

The concern that led me to this book is that environmentalism as a movement still appears to be primarily focusing on building an intellectual Maginot Line, rather than trying to establish a European Union. Polls show that the public in virtually every country, at every level of development, remains seriously concerned about their environment (although the particular issues that engage them tend to change with developmental status). But environmental treaties fail to be ratified by important countries, consumption and energy production in major developing countries continues to grow, and in the United States one of every two cars sold is a sport-utility vehicle (SUV) or a personal truck. More subtly, perhaps, the general environmental message increasingly seems dated: a vaguely countercultural critique of the modern world that in many cases is generally anticonsumption, antidevelopment, antitechnology, and anticapitalism, and is too often negative. There is a sense that, despite environmentalism having clearly become one of the most important strands of political and intellectual discourse in the latter decades of the twentieth century, a war is being lost. Why?

Could it be that environmentalism is fighting the wrong war, a war of the past, and ignoring the opportunities and challenges posed by new circumstances? Recognizing the cusp of change when a powerful movement has run its initial course and must mature and evolve, or fade into irrelevancy, is difficult, and hardly an objective task. But there are at least some signs that environmentalism, as a belief structure and ideology and as a critique of the rich and powerful status quo, may have reached—indeed, may have already passed—such a turning point. What are some of the indicators?

- Environmental law, which twenty years ago was a vibrant and actively changing area of practice, is regarded by many as a routinized backwater in the United States, where no new major environmental legislation has appeared for over a decade. Europe continues to innovate somewhat with product management and material bans, although the ultimate impact of such legislation beyond Europe is not clear. Thus, for example, the European opposition to genetically modified organisms (GMOs) does not appear to have stopped the evolution of that technology elsewhere.

- Environmental policy in many instances has ossified into a set of bumper stickers, each with its own institutional defenders. At the same time, rational solutions are increasingly rare—not because they do not exist, but because policy structures have become so adversarial and entrenched that the minimal levels of trust necessary for their implementation are impossible to achieve.
- Environmental science remains for too many an oxymoron, with a reputation for research intended to support whatever the respective activists or industry groups are arguing on a particular issue. Even some environmental scientists, conservation biologists, and industrial ecologists claim that their fields are valid only to the extent they support activist agendas.
- Environmental institutions—with some notable exceptions—are increasingly irrelevant: governments dither, firms generally continue business as usual, and most environmental nongovernmental organizations (NGOs) remain resolutely unaccountable. Generic public support for environmental issues remains high, but it is not quite clear what this means in a global economy where consumption continues to increase—as it must if developing countries are to continue developing. The sustainability dialog continues, but with less and less content and with a continuing tension between development and First World environmental demands.

These are gross generalizations, of course, and thus subject to the usual caveats, but taken together they indicate that environmentalism may indeed be at an important decision point in its history. A failure to evolve poses several potential dangers. It threatens the continued value of the environmental movement as an important critique of current market capitalism—and a healthy opposition is an important source of vitality for any evolving cultural system, especially a dominant one like capitalism. In fact, a significant strength of the Eurocentric, Enlightenment culture, and a major reason that it has grown to global scale, is precisely its ability to encourage such trenchant internal criticism; the loss of such a critique is not to be suffered lightly. Equally important, the underlying need to deal with a complex world increasingly dominated by human activity has not diminished—indeed, most data indicate that this need has become ever more pressing, not less.

How is such an impasse to be broken? My personal history leads me to an approach to this conundrum that differs from most. My intellectual and emotional involvement with environmental issues began, as it did for many, in the

1960s, when I studied the subject in college and on the streets (scented with tear gas as they sometimes were). I then detoured. After the army, law school, and an economics degree, I began working in telecommunications regulation. In retrospect, this was a significant shift for, although I remained sensitive to environmental issues, I was not continually embedded in the environmental worldview—in some ways, it was the equivalent of being abroad for a long time and realizing that one's native country and its culture were not the only reasonable patterns that humans could follow. Thus, when I returned to environmental issues as an environmental lawyer for a large company, it was with a broader perspective than some of my compatriots.

But environmental law was dissatisfying, because it was such an ineffective way of addressing the environmental issues I saw as increasingly important— loss of biodiversity, air and water quality, ozone depletion, global climate change, perturbation of the nitrogen and carbon cycles, and the like. Addressing such challenges had little to do with the Superfund and hazardous-waste statutes with which I worked. As a result, a few of us at AT&T, working with a small number of colleagues around the world, began to develop a new approach, industrial ecology, that drove environmental considerations back into technology design and institutional strategy decisions. This was not unique: while we worked with electronics firms to create methodologies such as design for environment (DFE), others at organizations like the Society for Environmental Toxicology and Chemistry were working on life cycle assessment (LCA) tools. Such efforts were not just responses to existing legal and public pressures; they were a profound recognition that environmental issues are perhaps more foundational than we thought.

Through industrial ecology and the random perturbations of corporate existence, I ended up working for AT&T as the vice president for environment and safety. I was also the champion of AT&T's telework/virtual offices and resulting workplace transformation program, perhaps the world's leading effort in that arena. In the former, I performed what most people would consider the corporate environmental function, but it was in the latter that I grew to appreciate a fundamental irony. The technology and cultural change that virtual organizations increasingly embodied—with their intranets, virtual publishing, reductions in unnecessary travel and resource consumption, and the like—had far more to do with environmental improvement than all the regulatory baggage taken together, yet the environment was never their main point. This became the case across industries: efficient routing algorithms for transportation and delivery firms significantly reduced the energy and environmental impact of such operations; digital photography in doctors' and dentists' offices eliminated their emissions of silver, an aquatic toxicant, from processing X-ray film. I came to realize that an environmentalism that is not as sophisticated about technological and cultural evolution, and about history, as it is about ecology or biology is in danger of becoming more and more like a Maginot Line: ineffectual, problematic, and anachronistic.

This need not happen. One can take the industrial ecology approach as a template that can be fruitfully applied at many levels of the global system, from the manufacturing facility, to the firm—to the level of global systems themselves. At this highest level, the approach becomes earth systems engineering and management (ESEM), a still nascent field that hopefully will, over time, move beyond the Maginot Line of environmental ideology to create new ways of perceiving and responding to the integrated, high-level challenges of a human earth. Like the EU project, ESEM is still very much a work in progress, and will undoubtedly be modified and improved as time passes. But new challenges require new solutions, and the evolution of a planet dominated as thoroughly as ours is by the activities of one species cannot be understood or managed just by using the remnants of old belief systems. This does not mean that environmental sensitivities or issues are passé—quite the opposite. But it does mean that, absent growth and evolution, environmentalism as a response risks becoming irrelevant, even harmful, to environmental progress.

The aim of this book, therefore, is threefold. First, I examine what has changed since environmentalism in its existing form burst forth in the early 1970s. Second, I then explored whether that approach is obsolete given the scale and properties of the challenges we currently face. In particular, the increasing importance of integrated human/natural systems at all scales, and the undeniable and growing impact of human activities on the dynamics of almost all natural systems, would appear to make any single approach—at least any we have come up with so far—oversimplistic and potentially misleading. Nonetheless, the outlines of a way forward can be discerned from these avenues of inquiry. Finally, I close with a discussion of earth systems engineering and management.

A Note on Form

This volume contains a number of essays first published in the *Green Business Letter* that, taken together, attempt to begin developing an environmentalism robust and realistic enough to address the urgent realities of our planet. The essay format has two limiting characteristics that might as well be addressed up front. First, it encourages challenging and provocative presentations, and, second, space limitations do not always allow sufficient room to provide thematic context. The former, of course, are an effort to encourage the reader to at least engage with the issues, and I apologize in advance if the result appears overvigorous or simplistic. Concerning the latter constraint, I have tried to create some integration in each chapter to compensate. In some ways, however, these drawbacks are less important given the subject area, for it is not at all clear that comprehensiveness is yet possible given our nascent state of knowledge. Indeed, the aphoristic format is apt in part because we are on the cusp of a transformation whose implications are still mainly hidden and thus cannot be fully explicated. As a byproduct of this format, I have also held footnotes and citations to a minimum; the book, after all, is not intended as an academic tome. For similar reasons, some of the columns have been slightly edited; the emphasis is on readability rather than archival accuracy.

I have sorted these essays into chapters reflecting major themes and trends. Each chapter opens with a brief introductory essay that outlines the main argument, followed by pertinent essays from the *Green Business Letter*, in which I try to crystallize a particular point in a way that I hope provokes the reader to further exploration, or students to classroom discussions. Next are is a list of references for the reader seeking more detail on particular issues.

The first two chapters, "The Human Earth" and "Real Rubber on Real Roads: Technology and Environment," address environmental issues from a different perspective than usual, one that arises not from the study of the natural world, but of the human world; and they introduce the concept of earth systems engineering and management. An immediate implication of this approach is a much stronger appreciation of the critical role of technology and technological evolution. Put simply, is it possible to understand environmental issues without understanding technology? This is, of course, a loaded question in that we do not really understand technology yet . . . and thus, logically, are that much further from being able to claim that we understand environmental issues. In asking this question, I am reflecting my experience in industry to some extent, where I came to appreciate the power of technology properly understood to change the very nature of environmental questions. Thus, preserving the rain forest translated to me as a challenge to implement corporate intranets and e-billing systems, thereby reducing unnecessary demand for paper. The next two chapters, "From Overhead to Strategic" and "Alice in Wonderland: Environmental Management in the Firm," fill in the specifics of this approach by discussing how environmentalism is evolving from a focus on very local and targeted interests, such as clean air and water or hazardous-waste sites in specific locations, to a far broader agenda, but in doing so is changing fundamentally in scope.

The next chapter, "Thoroughly Modern Marxist Utopianism: Sustainability," continues this line of thought by extending my critique to the cultural realm, in particular "sustainable development" and "sustainability." Such concepts, while useful if used carefully, in many cases merely enable evasion of understanding and the oversimplification of complex problems. Moreover, the extension of environmentalism to the social realm is both incomplete and profoundly challenging in ways that many have failed to grasp.

The next three chapters begin to explore this difficult terrain from different perspectives. "Faith and Science" attempts to understand why environmental science, particularly dealing with complex high-level systems such as climate and hydrological cycles and biodiversity, is regarded by many as problematic, and asks whether society is developing the basic information in these areas necessary for making informed decisions. "Complexity: The New Frontier" raises the challenge of complexity, an obvious challenge in trying to perceive, understand, and manage an anthropogenic (human-shaped) planet directly. We may lack not just the data and knowledge, but the mental models, concepts, and even the language necessary to responsibly operate in the world we have created. "How Humans Construct their Environment" discusses the cultural side of environmentalism, which is often inadequately recognized and many times buried underneath seemingly objective scientific language. It also discusses the potential for cultural imperialism and authoritarianism inherent in any ideology, and the inadequacy of such simplistic approaches in a complex world. In exploring these issues, the need for new ethical and, indeed, theological formulations appropriate to a rapidly evolving, increasingly human planet becomes clear, and some initial considerations are therefore proposed.

The final chapter, "Implementing Earth Systems Engineering and Management," offers some concluding thoughts. It is far too early to pretend that we can understand and easily respond to the challenges of ESEM. But this does not mean that we cannot develop a number of preliminary design principles, which can at least form a basis for future progress and provide some comfort that we are not embarked on an impossible mission. And, as always, one of the fundamental guidelines going forward should be not to let the best become the enemy of the good; not to let the daunting goals of understanding and managing our anthropogenic world stop us from doing much better now, even as we are learning.

As noted above, each chapter has a brief introduction of its main themes and arguments, and is followed by an annotated bibliography. I have included the bibliography because space limitations necessarily require that each chapter's essays simplify major intellectual trends and ideas, and thus to my chagrin limit my ability to guide the interested reader to works that could continue the intellectual dialog. These are not the usual books about industry or human society being gardens or boreal forests, or extolling sustainability as the egalitarian end point of human existence, many of which are both well-known to environmental audiences and frequently lacking in intellectual rigor. Rather, they are books that delve into the critical areas of knowledge that have yet to be sufficiently integrated into environmentalism—postmodernism, urban systems studies, evolution and function of technological systems, cultural anthropology and sociology, complexity and network theory, and, yes, philosophy and the like. That I have included a book in a chapter's annotated bibliography does not mean I agree with it—but it does mean that I was forced to think in order to disagree with it. In some ways, that is more valuable. By including them, I hope to provide the means for those who seek further dialog, for a discussion with a good book, even one you disagree with, is a powerful path to wisdom.

Certainly you may disagree with my points, or the facility with which I defend them. But I hope that in doing so you will at least be stimulated to reach higher ground where I may have faltered.

CHAPTER 1

The Human Earth

The Earth has become an anthropogenic planet. The dynamics of most natural systems—biological, chemical, and physical—are increasingly affected by the activities of one species, ours. The debate over how to manage global climate change; the efforts to recreate the Everglades and similar regional resource regimes to support both environmental and economic values; the effect of rapidly growing urban areas around the world on their hinterlands; the evolution of a global economy and market-oriented culture networked by information and communication systems that did not exist twenty years ago all testify to a planet whose characteristics, from the biological to the electromagnetic frequencies it radiates to space, are increasingly defined by human action. As the journal *Nature* put it in a 2003 editorial, "Welcome to the Anthropocene"—welcome to the Age of Humans.[1]

This trend is not new, of course. Although this process has been accelerated by the Industrial Revolution, "natural" and human systems at all scales have in fact been affecting each other, and coevolving, for millennia, and they are now more tightly coupled than ever. Copper production during the Sung dynasty, as well as in Athens and the Roman Republic and Empire, are reflected in deposition levels in Greenland ice; and lead production in ancient Athens, Rome, and medieval Europe is reflected in increases in lead concentration in the sediments of Swedish lakes. The buildup of carbon dioxide in the atmosphere began not with the post–World War II growth in consumption of fossil fuel, but with the growth of agriculture in, and thus deforestation of, Europe, Africa, and Asia over the past millennia. Humanity's impacts on biota, both

---

[1] *Nature* 424:709.

directly through predation and indirectly through the introduction of new species to indigenous habitats, has been going on for centuries as well.

What is different is that the impacts of the past were relatively minor and localized. Since the Industrial Revolution, they have become progressively more global and systematic (see "History, Responsibility, Design"). Indeed, in some areas, such as biotechnology, new fundamental technological and scientific advances have generated the potential for subsuming large chunks of previously (relatively) natural systems into human systems—in this case, genetic engineering combined with existing legal treatments of intellectual property creates the means by which the vast information store of biological genomes can be "commoditized" and made part of human economic systems (see "The Commoditization of Nature"). From a systems perspective, population and economic growth, and the evolution of ever more dense information and communication technology (ICT) infrastructures, has facilitated the linking of previously disparate local and regional patterns of human activity into globally synchronized systems, as well as ever closer coupling of human with fundamental natural systems. For the most part, we neither perceive nor recognize this state, in part perhaps because it has evolved without our conscious guidance; in part because recognition would require that we try to respond responsibly, rationally, and ethically and we do not know how to do that; and in part because the reality conflicts so basically with the popular mythology of "nature" as sacred. After all, to the extent we regard nature as something "outside the human," it becomes that much harder to recognize how much the human has actually affected the natural world.

So we should not be surprised that the language and mental models we often rely on in thinking about environmental issues have a powerful ideological and religious content, yet they are regarded by most of us as representing objectively real phenomenon. Thus, for example, terms such as "nature," "wilderness," "sustainable development," and even "environment" tend to be used as if they represented unquestionable, concrete facts and components of objective reality, but they are in fact products of a particular place, time, and culture, and have changed significantly over time. The concepts and meanings of "nature" alone make it one of the most complex of these cultural constructs. It can mean desirable; morally right; an independent source of value; the sacred; and, especially these days, the nonhuman, as "natural" ingredients are taken to mean "not made by humans" despite the fact that the products identified by such adjectives clearly are packaged, processed, and distributed in highly complex human structures. This implies that humans and their products, and especially their technologies, are somehow beyond the pale, profoundly "nonnatural."[2]

The irony, of course, is that this implication arises at the very point in human history characterized by increasing globalization of economic, technological, and cultural patterns. This does not mean that we are entering an era of global monoculture, but, rather, increasing complexity: there are more communities, units, systems, interests, political and social entities, and technology clusters, at many different levels, and more relationships among them and natural systems at many scales than ever before. Production, consumption, economics, culture—and environmental issues—are all part of the same, increasingly complex package, and viable solutions must deal with all the varied dimensions of that package.

Dealing ethically and rationally in such a world requires a sense of realism. Environmental issues are occasionally framed in apocalyptic terms, with people speaking glibly of "saving the planet." It is highly unlikely, however, that "the world," "life," or even the existence of the human species is threatened by current levels of human activity (indeed, as regards the latter, it is well-known that generalist species, of which humans are the outstanding example, do better than others in periods of rapid change). What is threatened is the stability of global economic and social systems; especially endangered, as always, are the poor and powerless, and those who do not belong to the dominant culture. Regional and global current system states—climate and oceanic circulation systems, biological systems at all scales, elemental and hydrological cycles, and the like—are also evolving rapidly. What the dramatic language indicates is not that the planet itself is threatened, for it will certainly continue to evolve, although elements of it will follow a different path than they would in the absence of humans. Rather, it is people's judgments about the world they want, and their underlying mental models and ideologies, that are at risk: it is not the planet, but individual, culturally contingent, and particular visions of a desirable state, that are under threat. And, of course, visions, and values, differ significantly depending on culture, economic status, and the like. Again, this is not to suggest that human activities might not result in sudden and unanticipated shifts in various critical systems—the climate or oceanic circulation systems, for example. But the widespread use of such apocalyptic language does suggest that the environmental discourse—the dialog, ideas, and institutions that taken together form the environmental movement—tends to conflate values, their vision of ideal ultimate ends, and reality with dangerous naïveté. What a "desirable" world is depends greatly on who, or what, is doing the desiring.

---

[2] W. Cronon, *Uncommon Ground: Rethinking the Human Place in Nature* (New York: W. W. Norton and Co., 1995) is an excellent source for those wishing to further explore these interesting intellectual byways.

This cultural contingency only adds to the high degree of complexity surrounding environmental issues. Consider scale, an important component of that complexity. The most common dimensions are geographic and temporal; even in these familiar waters, however, confusion arising from a lack of sensitivity to scale issues is all too common. Geographically, many scale issues arise because of a mismatch between the physical extent of human systems—jurisdictional boundaries associated with political systems such as nations, for example, or, more subtly, the geographic dimensions of markets and trading systems—and natural systems.

The temporal scale is a little more subtle. In the short term, social, cultural, and legal systems can reasonably be regarded as fixed, and policy development is relatively easy. Moreover, human individual and institutional perception is oriented to the short term: most people do not think beyond a few years. With the evolution of environmental focus from impacts localized in time and space, such as clean air, clean water, and defined waste sites, to perturbations that express themselves only over many decades, if not centuries, and over continental, if not global, geographical scales, the situation changes dramatically. Cultural constructs, fundamental institutions such as the family unit, the state, private firms and NGOs, and economic systems change profoundly, and none can be regarded as fixed. It is one thing to prescribe a scrubber technology that controls emissions of chlorinated solvent fumes from a manufacturing facility; quite another to mandate reduced fossil fuel use in response to global climate change. The former decision can be easily reversed if wrong; the latter significantly bounds the paths available for continued human evolution. It is not that a particular solution is "right" or "wrong"—rather, it is that scale issues make the latter a much, much more complex question.

Moreover, it is not clear that meaningful dialog on fundamental environmental perturbations, such as loss of biodiversity, global climate change, anthropogenic shifts in elemental and hydrological cycles, and the like involves only two dimensions. Indeed, one can argue that the appropriate "phase space"—set of dimensions required to functionally define the perturbations at issue—include a cultural dimension, a technological dimension, an economic dimension, and perhaps an information dimension as well.[3] And the choice of phase space may itself reflect the level of the system that one is interested in.

---

[3] Ideally, these dimensions should be independent of each other, but that is obviously not feasible with human systems such as culture and technology—nor, indeed, given the impact that human systems are now having on physical, chemical, and biological systems at all scales, may it be possible even in the temporal and spatial dimensions. Learning how to define, and perhaps quantify, such dimensions is an interesting challenge.

Choosing information technology as an example, there is a significant increase in complexity as one moves from a component, to a subassembly, to a router, to the Internet as a physical network, to the Internet as it functions in society, to the patterns of cultural change induced by the Internet as they in turn affect environmental systems. However, how to define each level in terms of complexity, and the way each level interacts with other networks reflecting other dimensions of the phase space at the appropriate scale is, to say the least, still opaque.

The challenge of complexity is significantly complicated in the real world by fundamental and accelerating changes in governance systems in a globalizing culture. "Governance"—the multitude of ways in which society is managed and administered—includes but is not limited to governments, which are formal institutions creating and administrating laws and regulations, and maintaining civil order. In the past, "government" and "governance" were quite similar. Indeed, since the Peace of Westphalia in 1648, the traditional global governance structure has been based on the institution of the nation-state. Thus, for example, the negotiations about mitigating global climate change are conducted entirely by nation-states, although firms and environmental NGOs participate and lobby behind the scenes. But this international governance system has become much more complex over the past decade. Where the nation-state used to be dominant, it now is just one of many institutions involved in international governance. Private firms, NGOs, and communities of different kinds now increasingly share responsibility for international policy development and implementation, notably in the environmental, human rights, and financial arenas. We have yet to catch up with this new reality, and most of the relevant players remain somewhat unclear about their roles in the still evolving governance structures.

And it becomes more complicated yet, for modern institutions involved in governance —from major religions and cultural systems, to the family unit, to geographic and virtual (online) communities, to firms, NGOs and activist movements—all continue to rapidly evolve. The power and information relationships they embody, and participate in, are shifting rapidly as well. Under such circumstances, there are no firm points from which one can begin developing intellectual frameworks and policy systems; rather, all is in fundamental flux. Cultural systems, institutions, and individuals tend to find such an environment uncomfortable, and will accordingly try to deny the reality of change and cling to previously valid patterns. Taken another way, rapid change creates the ironic but powerful incentive to neither perceive, nor acknowledge, the change that is actually occurring, and to retreat to the ideologies and mental models of the past. Indeed, the rapid rate of change that characterizes the modern world is probably a significant reason for the upsurge in fundamentalism, from the Middle East to the American South. Such a retreat is also appealing with a discourse as emotionally potent and ideologically charged as environmentalism: would you rather have the task of trying to understand and manage a world that is at the verge of the Anthropocene or be the romantic, muscular, dressed-in-black ecodefense warrior pictured in Foreman and Haywood's 1993 deep green classic, *Ecodefense: A Field Guide to Monkeywrenching*?

Governance structures also matter because shifts in power do not necessarily result in public benefit. One advantage of a government-based governance structure is that governments, to a more or less effective degree, are entities that implicitly try to balance mutually exclusive goals and conflicting ontologies, distributing costs and benefits of various policies among disparate groups—they are the integrators of civil society. Firms and NGOs, on the contrary, are far more limited in scope: firms generally exist to make money, and NGOs reflect their own idiosyncratic ideologies and belief systems, often exclusively. Both, given the opportunity, can dominate policy formulation to the exclusion of other voices and interests. It is probable that, as the current fragmented governance structure continues to evolve, mechanisms will emerge that integrate the social, environmental, and economic as the nation-state used to. For now, that desirable result does not always prevail.

This is especially true as the approaches of industry, governments, and NGOs to environmental issues tend to lag economic and technological evolution, and thus run the risk of focusing, like the Maginot Line, on past challenges rather than on present realities. The environmentalism that developed in the 1970s tended to focus on managing the effects of manufacturing and industrial activity, and to a lesser extent on products, particularly chemicals. If one assumes that environmental perturbations arise primarily from such activity, it is reasonably fair to declare victory and turn to other issues. If, however, environmental perturbations arise from the accumulated economic, technological, and cultural evolution of a rapidly growing population, and the increasing dominance of planetary systems by one species, then our ignorance is, in fact, far more pervasive than we might at first think.

Two trends suggest that this may be the case. The first is the shift in developed economies from a primarily manufacturing to a primarily service economy: depending on how "services" are defined, they now constitute 70 to 80 percent of developed economies. Moreover, this is not a simple shift in production technologies, for it is aligned with a substantial increase in the importance of information as economic input and output—the rise of the so-called knowledge economy. An environmentalism still focused on manufacturing deals but weakly with the environmental costs, and opportunities, of the modern service economy. Second, and equally important, is the growing realization that it is impossible to understand the modern world, and the interactions of human and natural systems, unless one understands technology systems and their evolution. Here, also, we know surprisingly little: in fact, we lack a robust theoretical framework. But it is certainly likely that human society will be profoundly changed by the confluence of information and communication technology, biotechnology, cognitive sciences, and nanotechnology. Consider some potential examples. How will future generations increasingly integrated into information networks rather than outdoor wilderness activities regard "nature"? What happens when biodiversity becomes increasingly a function of genetic engineering rather than existing species preservation? What is the effect of creating systems that enable humans to control machines and mechanical contrivances at a distance, wirelessly (the U.S. military is allegedly working on planes that are directly and wirelessly integrated into the brains of pilots, so that the plane becomes just another appendage of the human pilot)? What happens when medical science and nanotechnology combine to create human lives averaging well over a hundred years? Yet how often are these subjects considered seriously as part of the environmental movement (as opposed to simple Luddite dismissal)? Indeed, when are they considered seriously in public discourse generally? It is not merely rhetorical to wonder what a high-technology, high-information-content, environmentally desirable society might look like, for it is where we are heading—only, there is the distinct risk that the environmental piece might simply be an accidental, unintended, unplanned, and unpleasant result of other systems evolving, rather than a thought out, desired, and important component of our future.

Taken together, these observations suggest a fundamental caution that it is well to emphasize from the beginning: our ignorance about the environment is profound. Moreover, it is not trivial, in the sense that a little more research in the right direction will suffice to overcome it. Rather, it is foundational: it is not just the knowledge and data that are lacking, but even the perception of (and perhaps the ability to perceive) the complexity that we have unleashed over the past centuries. As D. Michael notes in *Barriers and Bridges*, referring to the experiences gained by those scientists trying to manage complex human-natural systems such as the Baltic Sea, the Everglades, the Great Lakes, and the like:

> Persons and organizations view information from their personal and peer-shared myths and boundaries. More informa-

tion provides an ever-larger pool out of which interested parties can fish differing positions on the history of what has led to current circumstances, on what is now happening, on what needs to be done, and on what the consequences will be. And more information often stimulates the creation of more options, resulting in the creation of still more information. . . . Indeed, in our current world situation, opening oneself or one's group to a larger "data base" reveals the terrifying prospect that the world is now so complex that no one really understands its dynamics and that even rational efforts tend to be washed out or misdirected by processes not understood and consequences not anticipated. Of course . . . those intent on pursuing their interests seldom can risk sociocultural ostracism by acknowledging this to others, and usually not even to themselves.[4]

In sum, it is not just the environmentalist discourse that sits uncomfortably on the cusp of unprecedented change. We cannot for much longer continue to evade the realization that we have created an extraordinarily complex anthropogenic world, one in which human systems and natural systems have become so intertwined that they are in many cases, if not yet in our minds and culture, one and the same. The carbon cycle, hydrologic cycle, climate and oceanic systems, biological communities—they all reflect the actions and intentions of one species. We are not yet prepared to manage this complexity—indeed, it is doubtful we understand either the systems, or what our response should be. But if we are to assume a position that is responsible, ethical, and rational, one that moves us beyond today's increasingly irrelevant ideologies and romantic fantasies, we must begin somewhere. We must develop the capability to design, through dialog and continual feedback, systems that achieve the multitudinous goals and desires of humanity, from personal fulfillment and economic security to environmental quality—an earth systems engineering and management (ESEM) capability.

Developing the capability to engineer and manage at the level of global systems—from energy, transportation, and information systems to the carbon and nitrogen cycles—is the next great challenge for our species (see "Earth Systems Engineering"). This ESEM capability is not a new burst of hubris, but an attempt to behave rationally and ethically in light of a basic, if disconcerting,

---

[4] D. N. Michael, "Barriers and Bridges to Learning in a Turbulent Human Ecology," in L.H. Gunderson, C.S. Holling, and S.S. Light, eds., *Barriers and Bridges to the Renewal of Ecosystems and Institutions* (New York: Columbia Univsersity Press, 1995.) pp. 461–488, at 473.

truth: the Earth, as it now exists, is a product of human design. So the question is not, as some might wish, whether we should begin ESEM, because we have been doing it for a long time, albeit unintentionally. The issue is whether we will assume the ethical responsibility to do ESEM rationally and responsibly.

So what is ESEM? It can be defined as the study and practice of designing and managing integrated human-natural systems in such a way as to provide the required functionality while supporting their resiliency and desirable characteristics over time. "Integrated human-natural systems" might range from the climate and carbon cycle, to the Everglades or Baltic, to large industrial systems producing commodity materials and energy, to urban systems. In each case, the relative importance of the human and the natural components of the system will vary, but rare is the system where both are not involved at this point. "Required functionality" is not limited to traditional human goals such as job creation, economic growth, or national security; for a system such as the Everglades, for example, it will include supporting high levels of biotic diversity and areas of reduced human presence. The latter are perfectly legitimate ESEM design goals; we are just not used to thinking of them in those terms. "Resiliency" is a critical property of a well-designed and well-managed system; especially at this early stage where our understanding of the relevant science and systems dynamics is liable to be limited, it is important that the systems we work with be relatively robust when the unexpected happens—as it frequently will. "Desirable" is a more difficult term than it might at first appear, but it marks an absolutely critical point: human values are determining how these systems are being structured currently, even if we often pretend they are not. They will continue to do so. Under these circumstances, whose values should dominate? How should different, often mutually exclusive, values be weighted and prioritized? Is your job worth my spotted owl—and who decides?

Given the complexity of these systems, it is apparent that minimizing the risk and scale of unplanned or undesirable perturbations is an obvious ESEM objective. But there is also an important caution here: the kind of engineering and management approaches that are almost universal today—which assume a high degree of knowledge and certainty about system behavior and a defined endpoint to the process—are not feasible in most ESEM situations. We will not "solve" or "control" in any traditional way ESEM systems such as the Everglades, the Baltic, the climate cycle, or the hydrologic cycle at a large scale. Rather, we will be in constant dialog with them, as they—and we and our cultures—change and coevolve together into the future. It is not our choice anymore as to whether our activities are integrated with these systems and determine their future evolutionary paths. It is only our choice as to whether we respond rationally and ethically to this challenge, or hide in outmoded and oversimplistic ideological systems (see "Climate Change and Social Engineering").

An obvious subject for the ESEM approach is the human relationship with climate systems, most recently and visibly through the global climate change research and negotiation process. While there are any number of criticisms of the Kyoto process, a few stand out. Some have to do with adequacy of governance: thus, for example, even though firms, trade groups, and NGOs are critical participants in any climate change policy initiative, the underlying structure of the negotiating process presupposes the traditional international governance system where states alone have sovereignty. Another fundamental problem with the Kyoto process is that it imposes a command-and-control, end-of-pipe mentality on a system where such thinking is manifestly inadequate. Even though a number of mechanisms are under discussion, Kyoto's principal focus is to limit emissions of carbon dioxide—in other words, to place a massive end-of-pipe control on the world's economy. This reflects traditional regulatory approaches, but it is grossly inadequate to both the problem and our responsibilities. Moreover, it assumes static technology and institutions just at the time that new technologies, such as active carbon sequestration at fossil fuel plants, offer the potential to achieve new ways of carbon-cycle stabilization.

The failure to appreciate the potential of technological evolution is fairly common, of course, but when one is dealing with policy structures that purport to extend over many decades, such a failure can render the entire exercise moot, for the power of technological evolution to radically redefine worldviews and unquestioned assumptions is evident. In the case of Kyoto, the negotiations began with the strong assumption that fossil fuel use must be reduced. As a second step, it is now being accepted by more moderate environmentalists that industrial carbon sequestration technologies, which capture carbon dioxide resulting from fossil fuel combustion in power plants and sequester it for centuries in deep aquifers, the ocean, geologic formations, or other reasonably long-term sinks, can enable fossil fuel use while reducing or eliminating associated emissions of global climate change gases. This is particularly important given the role coal may need to play in the development of large nations such as China or India.

But recent research indicates that it may be possible to scrub carbon dioxide from the atmosphere directly. This raises a significant cultural challenge, for it completely shifts the focus of climate change debate from "How do we stop affecting the climate" to "What kind of climate do we want?" The former question presupposes a world where human activities can be scaled back to the point where they have minimal impact; the second accepts a world where human demographics, economic and technological systems, and cultural evolution operate at a scale where simply withdrawing is a wistful utopian fantasy with the unfortunate effect of enabling a willful failure to accept ethical responsibility for the world as it is.

That we live in an anthropogenic world, and so obviously lack the necessary tools to do so rationally and ethically, may seem discouraging, both generally and more specifically for environmentalism. But this would be unduly pessimistic. A strong, responsible, and sophisticated environmental voice is necessary in shaping the world, and subsequent chapters will suggest some ways in which such a voice can be encouraged. And while it is true that ESEM is a new concept, we can already identify a number of principles from fields such as industrial ecology, adaptive management, and systems engineering that can be relied on to enable rapid progress in developing such an approach. A number of these principles are discussed in the last chapter. The Anthropocene is unquestionably a challenge, but—as this book itself shows—one to which we can begin to respond, always remembering not to let the best become the enemy of the good we can begin to do today.

History, Responsibility, Design
(October 2001)

One effect of living in the postmodern era is the attenuation of a sense of history, as events, ideas, and cultures from all ages are blended into an atemporal pastiche. This is problematic from a number of perspectives, but particularly from the environmental: by devaluing an understanding of the relationship between the human species and the Earth over the past millennia, it encourages two opposite, and oversimplistic, approaches. These extremes can be defined as, on the one hand, a belief that all environmental perturbations will be solved if the invisible hand of the market is simply allowed to work and, on the other hand, the deep green utopian vision of a depopulated world largely returned to "nature." Like most ideologies, these both demand a certain ignorance of history.

In this regard, a 2001 issue of *Science* that focused on "ecology through time" is worth noting, in particular its discussion of important ecosystems over time periods of decades to centuries.[5] Most people are aware of recent research indicating that ancient peoples had significant impacts on species extinction, especially megafaunal populations in, for example, Australia, North America, and Europe, and recent data indicating that human construction activity was widespread in areas such as the Amazon, which European explorers in their ignorance considered "pristine" and "natural." But it is of particular interest to read that "no coastal ecosystem now resembles its pristine precivilization state"[6] and that the collapse of fisheries, implicitly assumed by many to be a modern phenomenon, instead reflects centuries of human impact.[7]

A number of implications, most of which are challenging to the naïve and superficial view of environmental issues that tends to predominate these days, flow from this and similar studies. Thus, for example, it is becoming clearer that "primitive" peoples and societies were culturally no more sustainable than society today: where their technologies allowed, they also drove species extinct, harvested from natural resources such as fisheries in such a way as to drive fundamental ecological changes, perturbed local hydrologic regimes, and the like. What has changed is technology, human population numbers, wealth . . . but not the cultural (and perhaps innate?)

---

[5] *Science* 293 (July 27, 2001): 623–60.
[6] A. Sugden and R. Stone, "Filling Generation Gaps," 623. *Science* 293:623
[7] See J. B. C. Jackson et al. "Historical Overfishing and the Recent Collapse of Coastal Ecosystems," *Science* 293:629–37.

drivers behind the success of the human species. Indeed, the Jackson et al. study concludes that developing a "deep history" baseline, rather than one based just on recent experience, indicates that human impact on fish stocks and related biological communities has been going on for a far greater time than generally understood, even if such impacts are much accelerated with current population growth and evolution of fishing technology. Or, put another way, it is a far more anthropogenic world than we care to admit—or, for that matter, than our ideologies allow us to perceive. We comfort ourselves that we can just withdraw, and a pristine, prehuman "nature" can reassert itself. Data increasingly indicate that such views are wishful thinking, a product of ideology, not science. We cannot withdraw without substantial human mortality, and there is no "pristine" baseline to withdraw to.

Jackson and coauthors make another important point. The issue-by-issue, species-by-species approach that characterizes current environmental thinking is both profoundly unsystematic, and, more importantly, cannot begin to respond to the challenges of the anthropogenic world. Rather, such problems "need to be addressed by a series of bold experiments to test the success of integrated management for multiple goals on the scale of entire ecosystems."[8] In short, thinking of bits and pieces of "natural" systems without understanding their human context is in many cases highly problematic. It is not a particular fish on a Florida reef that is of concern: it is the massive design challenge of building a scientific, policy, and engineered infrastructure that simultaneously speaks to the needs of the Everglades; the Florida reef system as a whole; the increasing population and economic activity of Florida, especially on the coast; and agriculture, which in its turn is driven by a complex system of subsidy and political intrigue.

And thus the final point. Human impacts will continue to dominate in an anthropogenic world, and, accordingly, the ethical and rational response must involve design—and moral responsibility—at scales that have heretofore been beyond our practices and, indeed, capability. To deny the necessity for such earth systems engineering and management (ESEM) is not to protect what we have; it is to condemn it to continued degradation.

---

[8] Jackson et al., 636.

The Commoditization of Nature
(September 2000)

One of the most famous lines of *The Communist Manifesto* is Marx and Engle's reflection on the pace of change and secularization generated by bourgeois capitalism: "All that is solid melts into air, and all that is holy is profaned." This is a prescient observation, especially as regards the environment, for it targets an important dynamic, the importance of which is frequently unappreciated: the commoditization of nature.

"Commoditization" is a strange word. Frequently found in Marxist discourse, it means the process by which market capitalism changes things that were previously not regarded as economic goods into something with a price, and, concomitantly, into part of the economy. Thus, for example, a consultant may find that she is valuing her time based on her hourly charge, regardless of where it is spent: an hour watching TV or playing with her children costs two hundred dollars in forgone consulting income. Her time has been commoditized. In another example, critics of biotechnology and the patenting of genes argue that the process "commoditizes the genome," turning a fundamental natural system, the genetic material of a species, into a commercial good with a price attached to it.

Consider in this light the last part of the quote from the *Manifesto*: "and all that is holy is profaned." What this suggests is that elements of our life previously considered sacred are turned into economic goods by the action of the market economy. This dynamic—the pricing, and therefore commoditization, of the sacred that, by definition, is beyond price—is of particular interest in the environmental arena: genomes are commoditized; the right to pollute is commoditized; "nature" is purchased at specialty stores in malls, or experienced in parks run by large corporations. There are at least some environmentalists for whom, following the Enlightenment Romantics, nature has become the secular substitute for traditional religion. There are many more for whom nature in at least some of its manifestations, is sacred, beyond price. Yet without commoditization—call it internalizing the externalities—how are natural systems to be valued and preserved in a public discourse that frequently is defined by economics?

Perhaps nowhere are these conflicts more apparent than in the global climate change arena. "Emissions credits" are one mechanism by which efficient reduction of carbon dioxide emissions can be encouraged; they, in turn, are based on programs involving planting trees, preserving forests or grasslands, or the like. Such mechanisms, which rely on economic self-interest, may be crucial for efficient emissions reductions. On the other hand, it can be said with little exaggeration that the Kyoto process is historic in that it is the first time we humans have attempted to literally commoditize critical elements of a fundamental natural cycle—the carbon cycle—in a wholesale manner. What was once clearly exogenous to human culture is now being made endogenous: the carbon cycle, like the sulfur cycle, like the genomes of various species, is well on its way to becoming just another part of human economic activity. Marx and Engels would understand this well.

This commoditization process bothers many people on ideological and religious grounds. It is the dark side of the concept that the environment—"nature"—benefits when externalities are internalized. But one should be clear on what fundamentally drives this process: it is not some nefarious scheme by markets to dominate natural systems. Rather, it reflects the reality that the world is increasingly a human artifact, a monoculture reflecting the activities of one species—ours. In short, commoditization is not the means by which an otherwise unachievable control of natural systems is obtained, but a reflection of the fact that such influence over the dynamics of natural systems has already occurred. It follows, not precedes, the event itself. Commoditization of natural systems arises from the Industrial Revolution with its concomitant explosive growth in the scale of human economic activity and population levels, and the evolution of technological and cultural systems that accompanied it. It is a symptom, not a cause, of a new relationship between humans and their planet.

Because it reflects this reality, it is hard to see how commoditization of nature can be reversed without reversing the underlying structure—that is, without dramatic decreases in levels of human economic activity or populations. This is obviously somewhat problematic. Alternatively, we can work toward a governance system and ethical structure based on transparency, dialog, and multicultural collaboration, reflecting the realities of a human Earth—the direction of earth systems engineering and management (ESEM). Thus does the Marxist critique continue to call forth creative responses from the capitalistic, market-based system.

Earth Systems Engineering
(December 1999)

Last month this column discussed the need to apply an industrial ecology approach not just within the firm, but to broader human/natural systems as well. In particular, the need to move away from a traditional end-of-pipe model of global climate change mitigation and toward active carbon cycle management was highlighted. But the implications of earth systems engineering for environmental and business professionals are much broader than that.

To start with the big picture, it is important to recognize the systemic meaning of the Industrial Revolution and concomitant changes in our population levels, industrial and agricultural activities, technology systems, and culture. The result is a world in which the dynamics of major natural systems—carbon, nitrogen, hydrologic, sulfur, and heavy metal cycles; ocean and atmospheric patterns; the biosphere at every level from genetic to ecosystem—are dominated by human activity. Frequently, as in the case of invasive species, the impacts may be both unintended and a result of cumulative individual decisions, rather than directly engineered, but the results are the same. Managing these perturbations, and the future evolution of tightly coupled human/natural systems, will require the development of a capability to rationally "engineer" them—not in the usual engineering sense of control and precise definition, but by developing new managerial and engineering approaches which become part of an ongoing process of developing sustainably.

This capability is earth systems engineering, the study and practice of engineering technology systems so as to facilitate the active management of the dynamics of coupled natural systems. The goal is not to try to "engineer" natural systems in the traditional sense (engineering as control); with complex systems, such an approach is doomed to failure. Rather, the intent is to assume responsibility for any perturbations of natural systems that result from technological or industrial initiatives, and try to "design" the perturbations, rather than simply letting them happen (or, worse yet, pretending that they will not).

It sounds daunting, but in some ways we are already experimenting with it. The demand by stakeholders that firms consider the triple bottom line (TBL)—integrated social, economic, and environmental performance—is a move, albeit nascent and primitive, toward an earth systems engineering ethic. How so? Consider the example of biotechnology firms and the powerful reaction against genetically modified organisms, or GMOs, in Europe. Many of the firms involved had already gone beyond the traditional role of considering themselves only as profit-oriented entities. Yet they approached the issue of introducing GMO products as if it were a relatively routine business operation. In fact, it was quite different: the role of the firm shifted from being a developer and producer of products to being the manager, in the full social sense, of the introduction of critical new technologies with significant social and ethical dimensions. The firm shifted from the usual forms of business management and traditional responsibilities to being an important part of what can be seen as earth systems engineering.

Similarly, telecommunications firms, long used to pricing and equity issues being managed in the context of heavily regulated national markets, are suddenly participants in an "information revolution" that is global, unpredictable, and largely unregulated, and which could have significant social dimensions. Among the most important issues that arise are the environmental impacts of innovations growing from the Internet, such as e-commerce, and the distributional impacts of rapid technological evolution and growth of the knowledge economy (will the information revolution increase or decrease social and economic inequality?). Or, take fossil fuel firms: it is apparent that using fossil fuels can have environmental impacts. It is also apparent that nation-states such as China and India are hardly likely to eschew development if it involves the use of fossil fuels (both countries have significant coal reserves). What is the responsibility of fossil fuel firms to implement carbon sequestration and clean combustion technologies? What is their responsibility to facilitate the creation of a hydrogen energy distribution system so that carbon can be captured at a central location, and clean energy in the form of hydrogen deployed for mobile and perhaps stationary uses?

That these questions are even being asked illustrates several things. First, the scale of modern industrial and economic activity has indeed grown to be commensurate with that of natural systems: earth systems engineering is occurring, and will continue to occur, whether we consciously try to do better or not. Second, this dynamic is changing the traditional roles and responsibilities of firms, and their relationships to technology and society itself. Third, firms that understand this will have a greater probability of success in an increasingly complex world; firms that fail to adapt, like all complex systems that fail their evolutionary tests, will fail.

Climate Change and Social Engineering
(January 2001)

The headline on the article reads "World Powers Trade Charges on Climate Talks' Failure." The collapse of climate change negotiations (which I will call "the Kyoto process") is blamed primarily on the United States, identified as "the world's top polluter"; a South African paper charges: "The United States remains obsessed with the idea that it can use the dollar to buy itself out of trouble." Much sound and fury, but, frankly, very little light. The Kyoto process proves itself to be confused—but also an important and useful learning step.

To begin with, it is necessary to separate out the ideology from the science. Most importantly, the setback in the Kyoto process cannot be read as a rejection of the science of global climate change: most new data and assessment support rather than challenge the strong hypothesis that anthropogenic global climate change is indeed occurring. The details of the impacts, however, remain generally unclear: regional projections are still uncertain, and important components of the carbon cycle have yet to be fully understood. Thus, for example, it is not yet known whether North America is, overall, a carbon sink or a carbon source (and the time cycles and mechanisms involved are not well understood). The statement that the United States is "the world's largest polluter" is thus ideological and not scientific in nature, reflecting only one side of a ledger, emissions, whose other side, sinks, is perhaps not even perceived.

But the oft-repeated charge does bring out the social engineering dimension of the Kyoto process. It can be said that, for at least some stakeholders, the Kyoto process became not a way to balance the carbon cycle to reduce global climate change, but a way to restructure society and its values, especially regarding the United States. The consumption patterns and wealth of the United States are viewed by many as immoral and unfair, and, accordingly, the Kyoto process became a way to force especially American consumers to change. Thus, for example, German Environment Minister Juergen Tritten commented that the talks failed "because industrial countries [read: "United States"] wanted to count too much their natural forests as a source of man-made reduction rather than actually cutting greenhouse gases." The negotiating focus became restructuring the American economy, rather than managing the carbon cycle. This approach was all the stronger because of its alignment with the ideology of environmentalism, particularly the Rousseauian return to nature and "less is better" elements. It was also noted somewhat cynically by some Americans that implementation of the European Union policies would differentially affect U.S. economic performance, along the lines of "if you can't compete economically, compete through other policies."

The validity of various moral arguments or rationales held by various stakeholders, is not, however, the issue. What is relevant are three learnings that can be drawn from this experience, which are broadly applicable to earth systems engineering and management (ESEM) activities generally. The first is the integration of natural and human systems that characterizes the human world: in this case, moral and equitable dimensions of human systems have intervened in a process to directly affect economic and technological behavior that, in turn, has clear couplings to carbon cycle dynamics and, thus, climate system dynamics. The Kyoto process becomes a mechanism by which the carbon cycle is changed to reflect human morality.

The second involves governance. The Kyoto process has been dominated by an environmentalist discourse (and, to a limited extent, an economic discourse) that is, at best, neutral toward a number of other important discourses: market capitalism, technology, and libertarianism, to name but a few. And yet the goal of the process was to restructure societies, economies, and natural systems. Again disregarding the subjective validity of the positions of various stakeholders, one can conclude that as a governance and process matter, when such an ESEM process is undertaken, communication must be open, democratic, and inclusive. If it is primarily limited to one ideology or discourse, as in the Kyoto case, the process is unlikely to be either robust or successful.

And thus the final lesson: the dysfunctionality of ideology when it grows to dominate an ESEM process. That which looks backward and oversimplifies, as ideology always does, will seldom be a useful guide to a future that is in many ways a completely new and highly complex challenge. The Kyoto process is not dead, but when it arises from this particular fire, it will hopefully be a different and more inclusive Phoenix.

Annotated Bibliography

There are a few books that anyone who seriously wants to understand environmental issues should read at some point in her or his career. Many of them do not deal with environmental themes, or do so in unfamiliar ways, and indeed this is part of their value, for they help illuminate the context surrounding the environmentalist discourse. Indeed, this is a problem with many of the usual "environmental" books, which are so much a part of the current environmental discourse that they cannot help anyone see beyond it; they are in some ways explorations of the past rather than anticipations of the future. So, think about . . .

1. Marx. A brilliant critic (and thus architect) of capitalism and a source for much that is implicit in environmentalist critiques of capitalism and industrial society. As Robert Heilbroner has observed, "We turn to Marx, therefore, not because he is infallible, but because he is inescapable. Everyone who wishes to pursue the kind of investigation that Marx opened up, finds Marx there ahead of him, and must therefore agree with or confute, expand or discard, explain or explain away the ideas that are his legacy."[9] *The Communist Manifesto* is accessible and covers a number of relevant themes; other writings are somewhat more opaque and cover a lot of interesting, though possibly not entirely relevant, ground (I would go with a Marx reader, rather than *Capital* or some of his other complete works if you explore further). But if you want to understand utopian sustainability, globalization, commoditization, and a number of other important trends, Marx is still a good place to start.

2. D. Harvey, *Justice, Nature and the Geography of Difference* (Oxford: Blackwell Publishers, 1996). A wonderful book on postmodernity, cultural and economic transformation, "nature" and "environment" as cultural constructs, and how and what justice and environmentalism mean in a socially constructed world. Some of the language and approach is quasi-Marxist ("Fordism" might throw you the first time, and there are dialectics galore), but bear with it: the book is brilliant and commonsensical at the same time.

3. J. Diamond, *Guns, Germs, and Steel* (New York: W. W. Norton and Co., 1997) and D. S. Landes, *The Wealth and Poverty of Nations* (New York: W. W. Norton and Co., 1998). Most people see these books as somehow contradictory (Diamond viewed as a leftist and Landes as a rightist), but I think they are best read as complementary. Neither is an "environmental" book, but both address an absolutely critical environmental question: why does the modern world look the way it does, with a globalized, Eurocentric Enlightenment culture? After all, if you do not understand at least a

---

[9] R. L. Heilbroner, *Marxism: For and Against* (New York: W. W. Norton and Co., 1980), 15.

little about why the world looks the way it does and what the critical dynamics are, how do you think you are going to change it?

4. J. R. McNeill, *Something New Under the Sun* (New York: W. W. Norton and Co., 2000) and B. L. Turner, W. C. Clark, R. W. Kates, J. F. Richards, J. T. Mathews, and W. B. Meyers, eds., *The Earth as Transformed by Human Action* (Cambridge: Cambridge University Press, 1990). Environmentalists, like many moderns, frequently lack a sense of history and, accordingly, are not very good at explaining what exactly has changed over time to elevate environmental issues to the level of concern they now occupy. The short answer is "scale," but that will not get you very far. These books will. McNeill's book, since it is not an edited volume, is more readable and uniform in quality; the Turner et al. volume is a good, if a little dated, source for detail on particular issues.

5. K. S. Robinson, *Red Mars*, 1993; *Green Mars*, 1994; and *Blue Mars*, 1996 (New York: Bantam Books). This series, like many science fiction books, deals with, among other things, the ethical implications of terraforming a planet. What is interesting is that there is a large and relatively sophisticated literature on the ethical and philosophic implications of terraforming other planets (especially Mars), but virtually no discussion of those implications in the case of the only planet which, so far as we know, has been terraformed in fact—the Earth. One is tempted to think that this may be because we are unable to admit even to ourselves the illicit knowledge that we are doing precisely that—perhaps because we have for so long regarded the Earth, and now nature, as sacred?

6. F. Berkes and C. Folke, eds., *Linking Social and Ecological Systems: Management Practices and Social Mechanisms for Building Resilience* (Cambridge: Cambridge University Press, 1998) and L. H. Gunderson, C. S. Holling, and S. S. Light, eds., *Barriers and Bridges to the Renewal of Ecosystems and Institutions* (New York: Columbia University Press, 1995). Both of these edited volumes discuss, and provide case studies of, a relatively new area of study and practice called "adaptive management." This is defined in Gunderson et al. as "ways for active adaptation and learning in dealing with uncertainty in the management of complex regional ecosystems" (ix). The themes of complexity, the integration of human and natural systems, the necessary coupling of natural and social science research and expertise, and the dysfunctionality of ideology come across clearly in these works, which should be a part of every environmentalist's store of intellectual capital. Can you imagine the difficulty of actually trying to create an Everglades system in Florida that meets the design objectives and constraints of the various, mutually opposed stakeholders? Good. Because that is the current state of the entire planet.

CHAPTER 2

Real Rubber on Real Roads: Technology and Environment

Technology is a wonderfully complicated topic, made more so with the emergence of the Internet and the information society that are increasingly characteristic of global human culture. Hence, some have created taxonomies of technology—from hardware, to software, to "orgware," to "socioware"—in an attempt to explain the term in light of this migration to ever more ethereal realms. Whether this is useful is perhaps in the eye of the beholder, who may conclude that "technology" is at least intuitive, as opposed to, say, "orgware." Joseph Pitt in his book *Thinking About Technology* provides a useful insight by linking technology to "work," defined as "*the deliberate design and manufacture of the means to manipulate the environment to meet humanity's changing needs and goals.*"[1] This is a broad yet relatively simple formulation that we can modify slightly for our purposes to say that technology is the designed and built means by which humans interact with and affect their environment.

When technology is defined in terms of intentional design and construction, ethical responsibility for the results generally follows. There is, however, a major and important caveat in this fairly traditional approach. It is perfectly reasonable to assign moral responsibility to the technologist for a specific design and its resulting performance. But technology feeds upon itself, as advances in one area spark advances in other areas, and the resulting complex system restructures itself according not just to the plans of designers, but to its

---

[1] J. Pitt, *Thinking About Technology* (New York: Seven Bridges Press, 2000), 30–31, emphasis in original.

31

own internal dynamics and the social and cultural context within which it evolves—thus, students of technology recognize that such systems are autocatalytic and self-organizing. At this scale the question of who, or what institutions, are ethically responsible for technology systems is less easy. For example, if an engineer designs a router for use in the Internet, it strikes most people as fair to require a design that works, and does so safely and efficiently, as a matter of professional ethics. But the Internet itself is also a technological system, and one that is clearly all made by humans. It is far less clear that any individual engineer should bear the ethical responsibility for this system, which is far beyond any individual design decisions and whose present and future dynamics, even structure, are unknowable. Similarly, the designer of the first Portuguese caravel, the sailing ship that laid the technological groundwork for European global expansion, was no doubt held responsible for his ship design, but few would hold him responsible for the eventual results of European global colonization. The microethics at the personal level are fairly clear, at least conceptually; the macroethical level less so.

Even if a rigorous definition is difficult, however, it is possible based on the above observations to identify several themes that are absolutely critical for any consideration of technology and the environment. The most fundamental theme is that to be human is to be *Homo faber*, the tool maker. Understanding technology, in other words, is essential to understanding the human experience; in fact, technology is foundational to what it means to be human. Concomitantly, technology is also then the means by which environmental perturbations are introduced by human activity. And, contrary to the naïve view of some, this process is by no means strictly modern. The beginning of toolmaking was the beginning of technology—and recent scholarship suggests the beginning of widescale human impact on the environment, as tools combined with a nascent will to power[2] drove competing megafauna and some prey species extinct in Asia, Europe, and the Americas. Integrated technological, cultural, and demographic evolution led to settlements, the first built environments and primitive infrastructures for water, security, and protection against the elements, and to agriculture, perhaps the first systematic human technological transformation of the environment. Continued technological evolution can be read in many systems: Greenland ice deposits reflect copper production

---

[2] The "will to power" is Nietzsche's expansion in his 1901 *Wille zur Macht* of Schopenhauer's "will to live," and represents his recognition of the strength of the human drive to control and create (think of Goethe's Faust). More generally, one would certainly expect a priori that any species that grew to dominate a world would be characterized by a strong will to power, both over the environmental challenges it faced and within itself.

during the Sung dynasty in ancient China (around 1000 BC), and high concentrations of lead in sediments of Swedish lakes reflect production of that metal in ancient Athens, the Roman Empire, and medieval Europe. Air pollution began not in the 1950s in London, but with urbanization; Seneca in ancient Rome complained about "the stink, soot and heavy air" of that city. Anthropogenic emissions of carbon dioxide began not with the use of fossil fuels, but with the deforestation of Europe, North Africa, and parts of Asia a millennium ago.

This historic integration of technological, economic, cultural, political, and demographic evolution has led some more ideological environmentalists (sometimes called "deep greens") to argue that the "fall" of humanity as a species occurred with the beginnings of technology—of agriculture—ten thousand years ago. Since that time, the argument goes, technology and associated phenomenon—urbanization, widespread agriculture and irrigation, increases in population—have created a human cancer destroying the planet. Indeed, even many moderate environmentalists tend toward a reflexive skeptical attitude toward technology, as the debates over GMOs and industrial carbon sequestration show. But if technology in some sense defines the human, perhaps a more engaged posture toward its development (which need not be uncritical) might actually be more protective of natural systems.

Why does technology have this particularly complex relationship to society and cultural evolution? There are many reasons, but a major one is that unlike science, which (ideally) seeks to generate knowledge of present reality—"what is actually out there"—technology is more akin to art. Technology is a creative act, making certain evolutionary paths more possible and precluding others, positing a future vision of the world. It is not a coincidence that for the ancient Greeks and medieval Europe, arts, crafts, and technology were not differentiated: the Greek word for all of them was techne, meaning "art" or "artifice." Creating technology, like creating art, thus necessarily embeds choice and intentionality, and carries with it an implicit worldview. It reifies what was previously inchoate; more importantly, it reifies a specific and chosen set of options among a much larger universe of possibilities. For environmentalists, this tends to make technology problematic in several ways, because in essence the technological and environmental worldviews embed opposite teleologies, or visions, about the appropriate end state of the world. Environmentalism tends to favor a world of minimum human presence and impact; technology tends to favor a continually developing world characterized by ascendant human systems.

This already complicated relationship between the environmental and technological worldviews becomes even more so because of the reflexive relationship between the technological and the cultural state and the cultural constructs that are the basis of environmentalism. Technology is not just the structure through which culture and "natural" systems interact. More fundamentally, it is also a particularly powerful lens through which "nature," "the environment," "wilderness," and other constituent elements of the environmentalist discourse are perceived. The recognition that terms such as "nature" or "environment" are in part products of technological systems and cultures (think "nature" programs on television, or even "virtual reality" structures becoming popular in video gaming) is not an easy one for environmentalism (see "Look! In the Window! A Cultural Construct" and "Place and No Place: Environmentalism versus the Net"). What kind of world results, for example, if most children grow up in a mental world consisting of anime characters rather than live insects and animals, and think of evolution in terms of simulated cities and worlds rather than the one outside their window?

But the reflexive relationship between environmentalism and technology is not limited to the fuzzy and rather postmodernish arena of cultural constructions. Perhaps even more profoundly, the nature that for many environmentalists is the pristine "other" that must be preserved and protected is itself an artifact of earlier technologies and cultures (see "The Prickly Underbelly of Industrial Ecology"). That the world as we know it at this particular point in its development is primarily a result of human activity—technology—is difficult for many people to accept, but is increasingly apparent. Portuguese sailing technology of the fifteenth century, for example, changed the biology of virtually every island of any size, as it enabled the European colonization and exploration of the entire world. The process is an ongoing one: even today transportation technologies continue to affect biological evolution at all scales. Global shipping systems carry invasive species around the world, while modern aviation technology enables global tourism with its impact on fragile ecosystems in even previously inaccessible crannies of the world. Concomitantly, these transport networks create huge new target host pools for infectious agents, and thus become major instruments of viral and bacterial evolution. Similarly, recent work in the Amazon indicates that far from being the "natural" wilderness that Europeans took it to be upon their arrival—and a myth that many believe today—it has long been modified by humans, from construction of complex built systems, to modification of some 10 percent of Amazonia's soil by pre-European intensive agriculture (creating the so-called *terra preta*), to the existence throughout Amazonia of highly populated and relatively settled areas. In sum, it is technology that is ushering us into the Anthropocene.

But perhaps the most difficult characteristic of technology for many people, not just environmentalists, is its transformative effect. Consider the new technology described in chapter 1 that would simply "scrub" ambient carbon dioxide out of the atmosphere, thus allowing one to customize the atmosphere to whatever degree of greenhouse effect one wanted. If it actually works, the implications of such a technology are radical, for the question facing policy makers changes from "How can we stop using fossil fuels?" to "What kind of world do we want?" The first question raises no fundamental ethical question: it assumes that humans can withdraw from the world, at least in this instance, and that it is appropriate to do so. The second, however, is far more challenging, for it is a design question and thus carries with it ethical responsibility for the resulting choices. Moreover, the second arguably reflects the dynamics of technological societies more realistically: individual technologies, such as fossil fuel use, cannot be separated from larger questions of social and cultural structure, and overall technological competencies (see "Technology Systems and Salients"). Technological improvement and gains in economic, social, and environmental efficiency are indeed feasible and being made, but foundational change in technological systems, coupled as they are to other technological, economic, social, and cultural systems, is neither trivial nor easily accomplished. This point becomes increasingly critical as society, including environmentalists, begins to engage with the looming confluence of four major technological and scientific systems: nanotechnology, biotechnology, cognitive science, and information and communications technology (ICT).

The lack of sophistication about technology and technological evolution is widespread, and a better understanding of this area is something that the anthropogenic world demands of all of us. Indeed, it is almost axiomatic that, if we are to manage technology rationally and ethically, we need to understand it more comprehensively, and to do so at that point in its evolution when responsible intervention can be effective. Now, however, our ignorance is doubly dangerous: not only do we not know, but we don't know what we don't know. We thus make no effort to develop the knowledge even when we could. What, for example, are the environmental implications of the Internet, or of services such as e-commerce (see "'E-for-Environment' Commerce")? Now, when these systems are beginning to become embedded in global economic structures, is when we should be striving to develop systematic understanding of them. Similarly, technological change is occurring rapidly in the agricultural sector, yet systemic data and research on costs and benefits of various alternatives—high-technology agriculture, precision agriculture, use of GMOs, organic agriculture, no-till agriculture, integrated pest management, and the like—have not yet been comprehensively developed. How can we choose rationally among the alternatives, and determine where each is likely to be optimal, without such systemic knowledge?

Absolutist opposition to new technologies, from GMOs, to ambient carbon capture, to nanotechnology, is unlikely to be successful. The ideological certainty and lack of sophistication that such an approach embodies do not appear to halt technological evolution, but they do remove an important cautionary voice from the dialog about where and how such technology should evolve. Rather, the capability of modeling and evaluating technological systems at large scale, at a stage where they can be modified to reflect social and environmental concerns, needs to be enhanced. Particularly in an anthropogenic world, where such technologies are being implemented at historically unprecedented scales and speeds, it is important to begin addressing technological systems before they become embedded in economic and cultural practices and operations, and thus become far more difficult to modify (see "When Technologies Become Mythic"). Thus, a somewhat broad and speculative discussion of technology ends with a simple step that could produce significant benefits: encourage research on potential new technologies—the hydrogen economy, biomass, paper production from agricultural systems—before policies supporting their implementation are introduced (see "Technology and Environmentalism"). Although complex systems are in some ways constraining, humans are not powerless before technological change unless they choose, through lack of perception or oversimplification of the issues, to be so, nor is our only option opposition to all technological evolution. We may not know how to run in the Anthropocene yet—but we do know ways we could walk better, and more safely.

Equally important, a modern environmentalism must understand and respect technology, for otherwise it risks increasing irrelevancy. One does not need to be an apologist for technology to recognize its power, and the romance of the Luddite is no substitute for an ethical, if more sober, responsibility.

Look! In the Window! A Cultural Construct!
(February 2003)

It is fascinating to read about the history of the automobile and to realize that for many urban populations in the early decades of this century it represented a major means by which they realized "nature." Thus, for example, the 1925 Bronx River Parkway from New York City into Westchester County was originally conceived by Frederick Law Olmsted, who designed Central Park, as a "scenic utopia" reflecting an "authentic" American landscape.[3] It was planted with more than thirty thousand trees, each selected to represent those that were, or might have been, indigenous to the area. In this it was similar to Olmstead's Fens and Riverway project in Boston, which replaced acres of polluted mud flats with carefully engineered and designed "natural" patterns and plantings—or, for that matter, Central Park itself, a supremely engineered statement of what nature was for a city determined to ascend into the elite league of world cities.

Many interesting threads lead from these seemingly simple examples. One could, for example, examine how the parkway and other roads like it helped break down the physical boundaries of the city (from medieval wall to sprawl). One could contemplate "nature" as thus constructed from many angles: conservation of species; reconstruction of (highly idealized) landscapes; statements of power and class in a rapidly developing, immigrant-rich America. Thus, for example, Robert Moses designed low bridges over his parkway to Jones Beach on Long Island to prevent busses, and thus the lower classes, from accessing the getaway—although there was plenty of parking for the individual cars of the middle class. One could contemplate how each project explicitly redefined nature as an obviously designed and built environment at the time, yet came to be seen by future generations as natural in a foundational sense.

But another aspect caught my eye. Referring to the parkway and other contemporaneous efforts, Gandy comments that "nature became simultaneously more distant (framed by the window of a moving car), more accessible (through greater public contact with remote areas), and at the same time more individualized as an aesthetic experience."[4] In other words, the automobile as a technology system created a new and unique experience of nature—not the automobile alone, of course, for along with the parkway

---

[3] See M. Gandy, *Concrete and Clay* (Cambridge: MIT Press, 2002).
[4] Gandy, 122.

and other infrastructure was itself a representation of a certain moment in history and culture.

But consider nature through the window of the automobile—and then as it appears through the window of the television, more distant and yet, thanks to the power of nature photography, even closer, more emotionally evocative, and more accessible. Nothing is more artificial and contrived, more designed and planned, than the nature program on television—and more culturally potent, for that matter. Finally, there is nature as it begins to appear across the Internet, in at least three guises. The first is virtual reality: as bandwidth and processing power grow, nature will become entirely a human creation, whether in video games or created experiential landscapes of the future. The second is in complex models interpreting massive data sets: a nature that because it is "scientific" will appear more real than the former, but will be just as constructed and ideological. And just as removed, one might add, from whatever really is "out there." Third, consumers will increasingly buy their nature, from geode to cactus garden, not just from the museum or nature store at the mall, but over the Internet: nature as e-commerce.

This progression marks the increasing mediation of technological systems between whatever is "out there" and human perception and cognition; the evolution of a nature derived from human information structures rather than direct experience. It also marks the increasing commoditization of nature. The question of whether this evolution in the concept of "nature" is desirable is an open one; what is amazing, however, is that it is largely unseen.

For many, including environmental professionals and activists, "nature" remains a foundational reality, its character as a constantly changing cultural construct not even recognized. This greatly complicates environmental science and management, for, unfortunately, in the progression from perception, to analysis, to understanding, to wisdom, we have not yet really even begun to see. And thus the profound intellectual challenge posed to us by the Bronx River Parkway of 1925.

Place and No Place:
Environmentalism versus the Net
(March 2002)

Environmental issues are quintessentially a matter of a specific geographic place and time. A waste site exists over a certain period of time, occupies a specific place, and affects specific ecologies. The same is true of any environmental perturbation or, indeed, with species or biological communities. More generally, nothing seems more concrete—not just to environmentalists, but to almost anyone—than space and time. But not so fast: for a long time now, sociologists and philosophers have been focusing on two themes that undercut the foundational status of both space and time—and thus impact environmentalism as well.

The first generic theme, dating from at least Kant but made more recently prominent by intellectuals such as Barthes, is the critical role of language and signs in cognition. Put simply: culture is communication, and communication is signs and symbols. These signs and symbols combine in mental models or cultural constructs, which are the mechanisms by which any human intelligence perceives and understands the world. The second theme is quintessential postmodernism, backed by considerable sociological research: time and place are increasingly constructed pastiches, mixed together at will by institutions (think of Epcot Center[5]) or individuals surfing the Internet. As a result of modern transportation, communication, and information technology networks, what is increasingly real is what Manvel Castells in *The Rise of the Network Society* calls "the culture of real virtuality"[6]: what the television began the Net is accelerating. Time and space are increasingly categories by which information is filed, not boundaries on experience.

These themes can of course be exaggerated. But the sociological naïveté of many environmentalists (as well as others) has led such questions to be broadly overlooked. Thus, for example, it is apparent from even a cursory study of history that "nature," "environment," and "sustainability" are cultural constructs, reflecting the contingency of their time and place, and the

---

[5] Epcot Center combines in one place reconstructions of buildings and fragments of culture, such as foods, from many different places and historical periods. It is thus a prototypic postmodernist pastiche: one walks from medieval Europe to modern Asia in a few steps.

[6] M. Castells, *The Rise of the Network Society* (Oxford: Blackwell Publishers, 2000).

power of the elites that coined them. These categories are subject to change as culture changes.

That is nothing new. But add the postmodernist theme, and the potentially powerful effect of the Net, and interesting questions arise. Most obviously, cultural constructs change as culture changes, and studies indicate that developed-country cultures, whether in the United States, Japan, or France, are changing to reflect the impact of information systems and networks. These changes, to the extent they can be understood from our current vantage point, are telescoping time and place, rendering contingent what were (and are for environmentalists) absolute frameworks within which valued concepts are defined. Environmental activism requires time and place, and it is precisely time and place that are being made contingent by social and cultural evolution. If anything, the collapse of time and space as culturally foundational is accelerating—indeed, that is the fundamental reality of a profoundly multicultural world.

One need not accept this somewhat theoretical critique in its entirety to be concerned about the implications. Obviously, people will continue to be physical creatures located in particular space/time frameworks. But a culture and politics that increasingly devalues time and space as absolutes clearly has the potential to undercut virtually all the cultural constructs that today inform the environmentalist movement. Leave aside for the moment whether this is "good" or "bad"—indeed, in most cases it will be regarded as "bad" simply because those holding certain cultural constructs do not like to see them changed (or even challenged). Ask the more basic questions: what are the aspects of environmentalism that can be regarded as contingent and that can be discarded as history continues its ateleological evolution, and what elements, what perturbations, must not be overlooked? Concomitantly, what does it mean to be an environmentalist in a profoundly multicultural, postmodern world? What is the relationship between the environmental and the social?

You will not find answers to those kinds of questions in a 750-word column. Nor will you find them in most environmental policy or science programs. Rather, you must look for yourself. If you want to understand sustainability, begin with the study of Rousseau, Voltaire, and Marx. If you want to understand time and place compression, and pastiche, study the post-

---

[7] D. Harvey, *Justice, Nature and the Geography of Difference* (Cambridge, MA: Blackwell Publishers, 1996) and A. Giddens *The Consequences of Modernity* (Stanford, CA: Stanford University Press, 1990).

modernists and sociologists—David Harvey's *Justice, Nature and the Geography of Difference* , for example, or perhaps Anthony Giddens's *The Consequences of Modernity.*[7] You may well find that, if they are taken seriously, environmental issues are far more complex than even committed environmentalists understand.

The Prickly Underbelly of Industrial Ecology
(July 2002)

The term "industrial ecology" has long been recognized as an evocative analogy, suggesting the benefits of designing industrial systems to more closely resemble "natural" biological systems in their cycling of materials, energy, and waste. In some cases, there has been a failure to appreciate that the relationship between human and natural systems is one of analogy, not exact correspondence. This has led to interesting if questionable proposals to treat industrial structures and institutions as if they were gardens or forests. Thus, while industrial ecology is a powerful way to suggest new patterns of operations, it can be counterproductive when it leads to superficial commentary that fails to appreciate the profound differences between the two types of systems.

Human systems—including economic and industrial systems—with their reflexivity, autocatalytic dynamics, contingency, and intentionality, are much more complex than nonhuman systems. The logical and perceptual fallacy involved in conflating human and nonhuman biological systems is perhaps ironically illustrated by the fact that most such commentary displays strong ideological roots—and ideology is quite characteristic of human systems and hard to find in a salt marsh or boreal ecosystem.

I wish to invert the usual order of things. Perhaps blinded by the power of this metaphor to suggest that industrial systems should resemble ecosystems, few recognize that the analogy may be powerful in reverse: it is equally interesting and provocative to understand that many natural ecosystems can only be understood with reference to industrial—more broadly, human cultural and economic—systems.

Thus, for example, if I wish to understand the ecological structure of the Florida Everglades, I must first understand the flow of money, trade policies, and political power that created and nourishes the state's sugarcane industry, as well as Florida's overall demographic profile over time. To understand the rapidly changing ecology of the Aral Sea, I must understand the agricultural and cultural patterns of the Soviet Union regarding cotton. To understand changes in the world's forests, I must understand the pulp, paper, and timber industries, local governments and institutions, and, at times, their suborning through corruption.

Notice that this is not simply a change in the variables I consider as I look at an ecosystem. The coupling of such systems is far more fundamental than that. In many cases their dynamics, which may have been determined for ages by such nonhuman factors as predator/prey interactions, nutrient limitations, and the like, are now increasingly determined by human systems. In short, many natural systems now have structures and dynamics that reflect the reflexivity, contingency, and autocatalytical character found in human systems. This is true of both fundamental cycles—think of the carbon, hydrologic, and climate cycles—and biological systems at all scales—think of genetic engineering and, on the other end of the spectrum, the effect on island ecologies of the Polynesian and European migrations. In short, it is not so much a world of "natural capitalism" as of "industrial natural systems." It follows that the study of population biology, systems ecology, industrial ecology, and the like should begin with a foundation of sociology, philosophy and history. An interesting outcome.

This reading of industrial ecology suggests a different way of conceptualizing the study of ecosystems and, more broadly, any natural system. Currently such studies tend to emphasize the system's nonhuman elements—mapping, for example, nutrient and energy flows, or species distribution and predation patterns. But perhaps we need to overlay (not replace) these analyses with other maps—the flows of money and information, demand and supply patterns, infrastructure systems, and cultural and demographic dynamics, for example.

More difficult, perhaps, will be the study of the ideological and sociological determinants of natural ecosystem structure—the United States has national parks, for example, in part because of its powerful mythology about the American West. That all ecosystems are increasingly industrial is also based on analogy—but it is increasingly apparent that it is a much more powerful analogy than we may want to recognize.

Technology Systems and Salients
(November 2003)

Students of technology sometimes speak of "salients" and "reverse salients." The idea is that technological evolution is an advancing front with many components, like an individual wave flowing up on a beach. Some parts of the wave may run ahead but eventually are held back by others (the "reverse salients"). Of course, unlike a wave on a beach, the technology wave keeps flowing forward. This systematic model can be applied to specific technologies, such as bicycles or telephones: Wiebe Bijker and coauthors in *The Social Construction of Technological Systems* take such an approach (see this chapter's annotated bibliography). It can also be applied broadly to suites of technologies such as those that, taken together, constitute the Industrial Revolution (David Landes in *The Unbound Prometheus* takes this approach, although he calls the process "challenge and response," rather like Arnold Toynbee, not "salients").[8]

This gives a very different idea of technology, technological evolution, and industrialized cultures than the usual linear approach. We tend to think of technology as artifacts—a car, a computer—or perhaps as techniques. The "salient" approach, though, envisions technology as an integrated system: a particular expression in a time and place of a complex set of coupled artifacts, production and consumption methods, material and energy systems, mental models, cultural constructs, institutions, and even ideologies. The Industrial Revolution, for example, was not just the spinning jenny (invented by James Hargraves around 1765). Rather, it involved new technologies across the entire system required to turn raw cotton into printed cloth: cleaning, carding, preparation of roving, spinning, weaving, bleaching, printing, and marketing and transporting the cloth. At different times, the technology of a particular activity might lag, creating a "reverse salient"—which was usually rapidly overcome because of the strong economic pressures caused by progress in the coupled technologies. Today's example might be the coupled evolution of computer software, hardware, and networks.

Moreover, technological evolution is not easily bounded. For example, the evolution of technologies in the textile sector could not have occurred without (and was a causal factor in) the breakdown of the older, "putting

---

[8] D. Landes, *The Unbound Prometheus* 2nd edition (Cambridge: Cambridge University Press, 2003).

out" model of production (where home-based work was the norm) and its replacement first by small shops, then by large factories. It was also a factor in generating demand for new materials and sources of energy, so that factories could be sited without regard for waterpower—and the system thus included the industrialization of iron production and the shift from wood to coal. Among the results were the well-known demographic shifts from rural to urban work patterns, the rise of new institutions, and the eventual globalization of Eurocentric culture.

What does this have to do with environmentalism? To return to our example, within a few years the jenny caused civil disobedience—riots, in this case, rather than destroying crops, but the parallels with biotechnology are at least suggestive. Similar patterns may be emerging with nanotechnology, a technology system that has yet to be really defined, much less commercialized, but that has already been attacked by environmentalists, and with aquaculture, a very young technology that has already been dismissed as unsustainable by some deep greens. History and the "salients" analysis, however, suggest that such absolute opposition to specific technologies is both tactically and strategically unwise.

Tactically, because absolutist opposition to fairly specific technologies, while emotionally satisfying, overlooks the coupled nature of technological systems, bound not just to other technologies, but to economics, culture, and ideology. These technologies arise not in a vacuum, but as a part of an evolving pattern (in our case, a much more intensive focus of technology, economics, and culture on information structures), and opposition to a particular salient is highly unlikely to be successful. It may well, however, preclude rational and desirable shifts in the way a particular technology evolves.

Strategically, because the world as it is now, and as it must be to support six billion plus people, is a technological world. At its highest level, the system is heavily coupled, and even major cultural and technological trends are themselves salients. Trying to pick and chose technologies profoundly misunderstands the nature of technological and cultural evolution. That does not mean that principled, ethical individual and group action is not feasible. But it does mean that, absent a more sophisticated sense of the evolutionary patterns and drivers of these systems, any effect is likely to be minimal. Not raw emotion, but rather a less superficial understanding of the processes underlying technological evolution is a key part of living in a responsible and rational manner on an anthropogenic Earth—our true challenge.

"E-for-Environment" Commerce
(August 1999)

The great Austrian economist Joseph A. Schumpeter famously wrote of the "gale of creative destruction" that was capitalism; technological and institutional waves of innovation that broke upon the status quo, destroyed it, and built anew in a never ending process. E-commerce—the use of new information technologies to facilitate commercial activity—is such a pattern of innovation. Most people think of e-commerce in terms of online purchasing, or so-called business-to-consumer commerce, but this activity is fairly small compared to business-to-business commerce. In 1998, for example, consumer e-commerce was worth perhaps $8 billion, while intrabusiness e-commerce was around $43 billion; projections by the consultancy Forrester predict equivalent values in 2003 of $108 billion and $1.3 trillion.[9] Even at this early stage of the e-commerce gale, it is apparent that not just economic flows will change: customers will change their purchasing patterns; business cultures and organizational structures will evolve, perhaps dramatically.

So what? This is an environmental column, after all: what does e-commerce have to do with that? And therein lies the difficult paradox of modern environmentalism: the more important a technology is for the environment, the less it looks like anything having to do with traditional environmental activity. Discontinuous improvements in the environmental performance of a developed economy will not come from environmentalists, or environmental scientists, or environmental regulators, or environmental professionals: they will come from the operations of Schumpeter's gale. And how poorly equipped we are as a society to understand or encourage this.

Let us consider e-commerce from an environmental perspective. The most obvious point is that we have no analytical structure or methodology by which to begin evaluating the environmental implications of such a complex phenomenon. For example, e-commerce offers the potential of significantly improving the environmental efficiency of the economy. Each individual consumer does not have to drive their personal vehicle to a mall; rather, the fleets of mail and parcel delivery services run along their efficient, preplanned routes. Is this environmentally good or bad overall? Probably good, but no one really knows: it depends in part on things like the transport modality—air, truck, rail—chosen for various stages of the product

---

[9] Figures are taken from "Business and the Internet," *The Economist*, June 26, 1999, center section.

distribution system. It also depends on the efficiency of the routing algorithm used by the delivery service: in this regard, it could be said that the advances in mathematics that have enabled more efficient routing of vehicles among numerous points are possibly one of the most potent environmental technologies of the past decade. This is not a technology normally recognized by environmentalists and environmental regulators.

E-commerce can also cut waste significantly. For example, it enables companies to shift from offering products to offering services, with concomitant dematerialization of economic activity. Consider Home Depot, which has shifted from simply selling stuff to small contractors—its most important customer segment—to offering a Web site service. Contractors log in, enter details of their job, and the Home Depot software calculates what they will need and arranges for just-in-time delivery to the job site, eliminating the usual industry practice of overestimating materials, which then get wasted. What is ordered gets used. Or consider the new practice of printing books on demand: each book has a customer waiting, eliminating huge amounts of paper and energy that would otherwise be used to print and distribute books that never get sold, and end up being returned to, and discarded by, the publisher.

On the other hand, might not e-commerce stimulate consumption? Would people buy more in a primarily electronic world than a mall world? And is a product distribution system based on centralized package delivery networks really more environmentally efficient than individual trips to the mall? Under what conditions? And, perhaps far more important, what are the potential social and environmental implications of the evolution of industrial and commercial practices, cultures, and institutions that e-commerce may encourage? Can policies—including minimal regulatory interference—be developed that can encourage evolution of the e-commerce system in environmentally and socially appropriate directions? And how could we possibly know a priori what such directions may be?

These questions are unanswerable now. But we do not have the luxury of complaining that they are too complex to try to understand. The simple truth is that e-commerce is here, and it has significant potential for discontinuous improvements in environmental efficiency across broad sectors of the economy. We have a professional and ethical responsibility to deal with that reality. The phenomenon of e-commerce is thus an excellent example of the complex class of real-world problems that will become the core of twenty-first century environmentalism. Are you ready for them?

When Technologies Become Mythic
(September 2003)

It is only human to seek simple solutions to complex problems—the proverbial silver bullet. This is not necessarily bad, so long as everyone involved
remembers that it is only a strategy and does not lose sight of the complicated problems in the background. Such an approach can, however, become
a significant problem if the possible solutions become mythic—that is, if
they begin to take on the air of salvation stories as opposed to simple options
that may, or may not, work in the real world.

Somewhat ironically, a few of these mythic solutions in the environmental area are technology systems, such as the suites of technologies collectively
summarized as "the hydrogen economy" and as "biomass." The former contemplates the widespread use of hydrogen as an energy source, especially for
mobile uses; the latter contemplates a substantially increased focus on biomass as a source of fuel, material, and functionality (e.g., part of constructed
urban infrastructures). Both are highly appealing politically and ideologically to many for environmental reasons. In both cases, however, this eleva-
tion to mythic status is potentially problematic.

Consider briefly the hydrogen economy. Most obviously, hydrogen, like
electricity, is a secondary energy source: it has to be generated from a primary fuel such as coal, petroleum, or nuclear. That hydrogen itself is environmentally preferable is, therefore, immaterial unless the generating
process is also considered. Second, hydrogen cannot be distributed through
existing infrastructure, and even with new infrastructure substantial leakage
is likely. This raises both efficiency and operating safety (and liability) issues.
It also suggests that hydrogen use at the scale contemplated could have—
indeed, must have at least at the local level—substantial effects on atmospheric chemistry. It is thus somewhat surprising, especially given governmental enthusiasm for hydrogen programs, that no systemic analysis
covering the technological, environmental, economic, and social end-to-end
implications of an economy based on hydrogen has been attempted. This is
no doubt due in part to the mythic status of the hydrogen economy; why
question that which is already known to be good?

Similarly, increased reliance on biomass, primarily for fuel and material,
is problematic at the scales implied by some enthusiasts. Serious land-use
and loss of biodiversity issues are involved in any transition of land to biomass production at scale. Moreover, the energy, fertilizers, and pesticides
necessary to support significantly increased biomass production have their

own impacts. For example, agricultural activity in many regions has already had significant impact on estuarine areas; the large dead zones in the Gulf of Mexico, Long Island Sound, and elsewhere are evidence of that. Increasing the scale of biomass production significantly will only worsen the perturbations of the nitrogen and phosphorous cycles that underlie them. The only viable solution may be genetic engineering, but that is intensely problematic for many environmentalists, even those strongly supporting biomass technologies.

The immediate problem regarding the mythic status of these, and similar, technological systems is that it leads to the assumption that they can substantially address major environmental and economic issues in a simple, permanent, and desirable manner. Thus, for example, the hope is that the hydrogen economy can provide energy without attendant environmental impacts. This may be true—that something is mythic does not mean that it is ineffectual. But the effect is to blind people to the need to carefully test and evaluate new technologies as they are introduced, and, especially at the scale of technologies such as these, to be constantly on guard against potential rebound effects. Every new major technological system should be assessed before deployment, at least to the best of our ability to do so, and especially in cases where such technologies appear on first blush to be highly desirable (and thus suppress critiques).

More fundamentally, however, the danger of mythic technologies is that they enable our refusal to deal with reality. The truth is that our species at its current scale of population and activity is beyond silver bullets—no technological systems, no matter how mythic, can provide permanent and simple solutions. Rather, we are now, and will be until our economic or demographic collapse, in constant dialog with global systems—the nitrogen and carbon cycles, the hydrologic and climate systems, biological systems at all scales—and solutions will be complex, will involve difficult trade-offs, and inevitably will be partial and contingent. Comfortable, even desperate, belief in salvation myths, be they technology or sea beast, only serve to blind us to our responsibility to respond ethically and rationally to the challenge of this anthropogenic Earth.

Technology and Environmentalism
(June, 2000)

If one were scripting the dance between technology and environment, it would be a strange performance indeed. Contrary to today's ahistorical postmodernism, this "dance" has existed since early humans began driving megafauna extinct in Australia, or smelting metal in China, Greece, or Rome, or deforesting Europe and North Africa between the tenth and fourteenth centuries. Nor are end-of-pipe command-and-control regulations a modern invention: as early as 1306, London adopted ordinances limiting the burning of coal for air-quality reasons (which did not prevent "killer smogs" in 1873, 1880, 1891, and 1952, the latter leading to new clean air legislation in 1956). Another example is the English Alkali Act of 1863, passed to control emissions of gaseous hydrochloric acid resulting from the LeBlanc method for producing soda (sodium carbonate). This law even imposed a form of "best available control technology" (BACT) in the form of acid absorption towers designed by William Gossage.

Such examples, and their modern analogs, illustrate the continuing dialog between environmentalism and technology. And, indeed, modern environmentalism is a powerful movement for addressing simple and easily observed problems. Clean air and clean water legislation, waste site cleanup, preservation of highly valued landscapes, protecting visible and sympathetic species from extinction, and even bans on materials whose problematic environmental impacts can be easily demonstrated (for example, lead in gasoline) are examples. The successes achieved by environmentalism in these cases are real and important.

Potential dangers arise when the tools and mental models appropriate to simple solutions are applied to complex ones, where they do not work. From a political perspective, surveys continually demonstrate strong support for environmentalism, but this primarily extends to easily seen problems—one of the reasons that much environmentalism focuses on manufacturing, where environmental insults are both easily detected and relatively easily fixed. When environmental issues involve complex systems with time cycles measured in decades or centuries rather than months, and where scientific uncertainty is relatively high, public support falls off rapidly. This explains to some extent why the American public is strongly supportive of environmentalism generally, but deeply split over the Kyoto process. This is an obvious complexity when attempting to craft long-term policies to address fundamental environmental perturbations.

A more basic problem, however, arises from the historical evolution of environmentalism, which has generally positioned environmental issues as "overhead"—that is, something to be taken care of only after primary missions are accomplished. In firms, this is represented by end-of-pipe technologies such as scrubbers or water treatment plants, which are relatively independent of product or process design. The mental model behind this approach, however, encourages simplistic, often ideological, approaches to very complex problems. It thus has problematic implications when applied to complex, real-world technological systems.

Consider, for example, the costs of a wrong decision in an end-of-pipe scenario: at most, it will result in a little wasted capital or an inadequate level of protection that, once recognized, can be fixed by a quick and simple switchout of the control technology. The penalties for being wrong are small and easily fixed. End-of-pipe technology, in other words, is a simple system. But this changes when core technologies, linked with other technological systems and embedded in a complex cultural and economic matrix, are mandated. The complexity of these technological systems is orders of magnitude greater than with control technologies and the costs of wrong decisions, to the economy and to the environment, can be far greater. Thus, for example, if rather than a scrubber one mandates a change in chemistry of gasoline, one affects not just a factory, but a complex technological system, and the costs of being wrong, as in the case of the required addition of MTBE to gasoline, can be billions of dollars and billions of gallons of unnecessarily contaminated groundwater.

This leads to what should be a simple rule for all: prior to encouraging a fundamental technological change, it should be standard practice to at least try to identify the real-world impacts that might result. This should be the case whether it is a firm, a government, or an NGO urging the change: it should apply to GMOs, to government policies encouraging biomass plantations (and thus possible increased distortion of the nitrogen cycle), and to NGO demands for bans of important industrial materials (what will replace them? And how can one know what is better unless one has tried to answer this question?). And, in doing so, the complexity of the real world, not the simplistic language of ideology, should be our guide.

Annotated Bibliography

1. A. Grubler, *Technology and Global Change* (Cambridge: Cambridge University Press, 1998). An excellent and rigorous discussion of the evolution of technology and how that process has generated global environmental change. The book is useful not only for its plentiful data, but also for its historical scope and its sophisticated understanding of the underlying processes of technological innovation and diffusion. It is among the few, and the most successful, in integrating two discourses, the technological and environmental, which usually—and unfortunately—continue to treat each other as irrelevant.

2. W. E. Bijker, T. P. Hughes, and T. Pinch, eds., *The Social Construction of Technological Systems* (Cambridge, MA: MIT Press, 1987). You cannot hope to understand technology if you do not understand the historically contingent creative process by which it arises and the nexus of economic, social, scientific, and other streams that come together in any successful technology. This volume is rightfully considered a classic in the history and sociology of technology—but don't let that put you off. It is a great education in itself.

3. M. Heidegger, original essays 1952–1962, collected in 1977, *The Question Concerning Technology and Other Essays*, translated by W. Lovitt (New York: Harper Torchbooks); W. Barrett, *The Illusion of Technique* (Garden City, NY: Anchor Books, 1978); and J. Pitt, *Thinking About Technology* (New York: Seven Bridges Press, 2000). Technology is the essence of the human, so it is not surprising that philosophers have struggled with the questions it raises for centuries. Heidegger is among the best and most trenchant of the recent critics of technology, although he is not an easy read and some of his concepts, such as the Tao-like Being, will throw the uninitiated. Barrett's book, by contrast, is both readable and solid, while Pitt's is perhaps a little more academic. Leaking through these treatments, however, is the disquieting notion that you cannot truly understand technology until you understand intentionality, free will, and the Nietzschean will to power as they function in the real world . . . subjects that have challenged philosophers for centuries.

4. D. F. Noble, *The Religion of Technology* (New York: Alfred A. Knopf, 1998). Noble's classic work traces the powerful integration of theology and the technological arts that uniquely marked medieval and Enlightenment Europe, a dynamic that is a major reason why today's globalized culture is Eurocentric. While a fun and fairly easy read taken alone, interested readers seeking further enlightenment should read Joseph Needham's 1956 classic study of Chinese science and technology, *Science and Civilization in*

China: Volume 2: History of Scientific Thought (reprint, Cambridge: Cambridge University Press, 1991) in juxtaposition. The latter is a more academic and researched tome, but the two taken together create the kind of intellectual experience you will treasure if you are truly interested in understanding the world we find ourselves in.

CHAPTER 3

From Overhead to Strategic

Environmentalism is not a new phenomenon—in ancient Rome, Seneca complained of the stink and soot in its air. In 1851 *Scientific American* described London in apocalyptic terms:

> The 300,000 houses of London are interspersed by a street surface averaging about 44 square yards per house, of which a large proportion is paved with granite. Upwards of two hundred thousand pairs of wheels, aided by a considerably larger number of iron-shod horses' feet, are constantly grinding this granite to powder; which is mixed with from 2 to 10 cartloads of horse-droppings per mile of street per diem, besides an unknown quantity of the sooty deposits from half a million smoking chimneys. The close, stable-like smell and flavor of the London air, the rapid soiling of our hands, our linen, the hangings of our rooms and the air-tubes of our lungs bear ample witness to the reality of this evil.[1]

But modern environmentalism as a movement can probably be dated to the publication of Rachael Carson's *Silent Spring* in 1962. A number of pressures combined to shape the nascent movement, including the general zeitgeist of the 1960s. Marxism, or at least socialism, was a much more widely accepted critique of capitalism than it is today, and its strength validated other critiques. There was also a widespread skepticism of technology that, in the guise of nuclear weapons, was seen as threatening the world itself. Technology in the form of factories was seen as destroying the world through pollution. These

---

[1] "50, 100, and 150 Years Ago," *Scientific American* 285, no. 1 (2001): 16.

trends aligned with and strengthened a belief, with roots in the European Enlightenment, that in an increasingly secular age "nature" was sacred. Perhaps most definitionally, environmentalism originated within and adopted a powerful emotional and ideological antiestablishment stance.

The antiestablishment bent reflected the countercultural attitudes prevailing among the youthful supporters of environmentalism, but also political dynamics: the political and industrial institutions of the day were not inclined toward environmental concern, and only such an emotionally and ideologically powerful movement could have had a hope of success. These pressures created an aggressive and assertive movement, one convinced that environmentalism was the source and main defender of foundational truths necessary to "save the planet." This romantic self-image has proven hard to dislodge, and, as it increasingly diverges from the public's perception and values, may in fact be a source of weakness (see "In the Beginning . . ." and "Values, Goods, and the Environment").

This stance, combined with a receptive public that understood the concerns about polluted air and water because they could see, touch, and smell it, was extremely successful in the short term. It did, however, have one major effect: it positioned "the environment" as a concern that everyone—environmentalists, environmental professionals, private industry, governing institutions—was comfortable treating as "overhead" rather than "strategic." In business terms, an overhead function is one that has nothing to do with the production function of the firm. A strategic activity, on the other hand, is one that is integral to the operations and existence of the institution. In societal terms, environmentalism has been seen as an overhead—a cost to society itself, one that has little to do with the business of ordering our social affairs and maintaining a sound economy and national defense. Among other things, this institutionalized environmentalism is seen as "a loyal opposition" rather than an important contributor to major operations and decisions across the board, critiquing from the outside rather than trying to balance the myriad conflicting interests facing the decision maker (see "A Government in Exile"). It also encouraged an approach that allows nonenvironmental institutions to treat environmental issues as tangential to their core missions and decision making. Indeed, this approach remains embedded to a large degree in educational institutions as well, which tend to teach environmental science, policy, and economics as separate, if rather confused, disciplines, instead of integrating environmental considerations as an important part of the context for most disciplines (see "But I Want to Work on Environmental Stuff").

Environmentalists and environmental professionals generally see this isolation from the main decision-making table as a bad thing and they strive to overcome it. It is not clear, however, that many of the movement's leaders truly understand what this evolution means for them. When the environment is considered overhead, it dominates decision making in its domain, though that domain is limited to environmental issues such as cleanups, emissions control technologies, and the like. Conversely, when the environment does become strategic, it may be considered as part of a larger domain, but it will not dominate the decision-making process. In technology, for example, if one is making a decision about what kind of air scrubber to purchase, environmental considerations are paramount, but if one is designing a nonenvironmental technology like a telephone or automobile, environmental considerations may be present, but they are seldom dominant (see "Green Technology: From Oxymoron to Null Set"). In fact, the goal should be "good technology," with the understanding that in today's world that implies that environmental and social considerations have been integrated into the design. When the environment is seen as overhead the decision process is internal to the environmental community and policy systems; when it is considered strategic, many other discourses may be involved and the need to compromise, and understand and respect other viewpoints, may be more demanding. This is more subtle than it looks: the training and psychology of a person comfortable with a tightly bounded, overhead position are quite different than that required for a person in a strategic position. Thus, for example, environmental professionals in industry, who have self-selected and been hired for their specific skills, frequently talk about becoming strategic, but they are generally unqualified for such positions, setting up a serious case of institutional cognitive dissonance (see "Implementing Strategic Environmentalism").

This dynamic does not only occur with technological issues or in industry. Consider the somewhat fractious dialog between the environmental and international trade communities. There are many details to this exchange, but the underlying dynamic is in some ways easy to discern. As environmental concerns evolve from merely overhead in international affairs—that is, a set of issues to be dealt with only by environmental ministers and limited to environmental policies—to strategic, they inevitably begin to affect policies and processes of other, powerful communities, such as the trade community. In doing so, the initial response of both communities tends to be defensive and adversarial: environmentalists assume they are right on all counts, and the trade community does the same. But with time, and perhaps a better understanding of the legitimacies that underlie each policy structure, what began as conflict can transform into dialog, compromise, and the evolution of integrated policy structures. Thus, environmental and national security concerns can be managed as "environmental security" concerns by elements of both

communities (in the case of the United States, by the State Department working with the Environmental Protection Agency's international group). It is in
the latter stage of integrated policy structures that the environment is truly
seen as strategic, but this results in a concomitant loss of absolute authority in
its domain that marked it in its overhead phase.

Interestingly, the overhead to strategic transition also occurred with technology itself during the Industrial Revolution, as David Landes notes in *The
Unbound Prometheus*:

> In a strange way, the importance of technology as a factor in
> economic change was thus both heightened and diminished [by
> the institutionalization of technological expertise in the transi-
> tion from overhead to strategic competence during the Indus-
> trial Revolution, especially the late 1800s]. On the one hand, it
> became more than ever the key to competitive success and
> growth. The faster the rate of change, the more important to be
> able to keep up with the pacemakers. On the other, technology
> was no longer a relatively autonomous determinant. Instead, it
> had become just another input, with a relatively elastic supply
> curve at that.[2]

And thus, in a nice historical irony, the environmental discourse in its turn follows the path of technology (and the study of the former offers insights into
how to manage the latter gracefully).

From a broader perspective, environmentalism like most human discourses
arises out of a particular view of what human beings should strive toward, what
the "right" endpoint of human existence is. To the extent that environmentalism
was only overhead, this question did not matter, for overhead functions seldom
challenge more strategic or fundamental belief systems. But when we begin to
see environmental issues as strategic—and they obviously are in an anthropogenic world—the implicit endpoint (or teleology) presupposed by the environmental worldview matters, for it is at that point that ideas about what constitutes ultimate value become the basis for decision making. Our policy choices
will differ depending on whether the final state we envision is a world with only
a few humans, otherwise left to evolve on its own, or a world with billions of
people on it supported by significant technological infrastructure.

Whether the former vision of Earth is one widely valued at this point in

---

[2] D. S. Landes, *The Unbound Prometheus: Technological Change and Industrial Development in Western Europe from 1750 to the Present* (Cambridge: Cambridge University
Press, 2003), 326.

human historical and cultural evolution is questionable. Put oversimplistically, while it is clear that for many people environmental considerations are important, there are no data that indicate that most people in developed countries are excited by the prospect of going "off the grid," giving up consumption, becoming vegan, and adopting a deep green lifestyle (see "Brand Image").

The implications of this are not clear. But it does appear as if the shift from overhead to strategic, from antiestablishment to establishment, carries with it an implicit redefinition of environmentalism and that so far the community has not made this shift. The anthropogenic world needs a strategic environmentalism, but so far we have not seen what that might look like.

———

In the Beginning . . .
(April 2003)

For many environmentalists, the current state of the world is problematic: too many people using too many resources and perturbing too many natural systems. Assuming this to be true—and some don't—there are two fundamental responses: reduce human activity dramatically or embrace an anthropogenic world. Many environmentalists incline toward the first option, leavened by political and moral reality. But I want to ask a more basic question: why does the world look the way it does, and can that shed any light on potential future direction?

Now, one clearly does not answer such a question in a short column. But obvious points can be made. First, uniquely in the known universe, one species has risen to dominance over a world. That does not mean ethical and rational control, but it does mean that, over the past ten or twenty thousand years, one species has learned how to capture significant amounts of the available energy and resources and convert much of the planet's environment to suit its own interests. Second, this is not unique to the modern era: anthropogenic changes in atmospheric carbon concentration began a thousand years ago with the deforestation of Europe and North Africa; lakes and fossil ice around the world show spikes in concentrations of lead and copper reflecting the rise and fall of civilizations in China, Rome, and Greece; new evidence indicates that "primitive" peoples significantly altered the Amazon to support large populations and modified nearshore fisheries and estuaries around the world to provide food; and manipulation of the genetics of other species for human purposes began with the advent of agriculture. Some believe that non-European civilizations had a more "natural" or "sustainable" culture—a view that frequently coincides with a New Age placement of the sacred within nature—but modern scholarship increasingly cannot support this displaced golden age utopianism. What has changed is the scale, and technological leverage, that humans now can bring to these traditional activities, not the inherent will to power. Concurrently, cultures compete with each other and accordingly evolve, both internally and taken as a whole.

This technological and cultural evolution, unique to the human species, has three important characteristics. First, it does not appear to be directed to any particular end, and whether it is "good" or "bad" depends on the historical and cultural perspective of the observer. Second, while perhaps not "progress" in a normative sense, it does exhibit a continuing tendency towards greater complexity. Third, the overall effect is to reinforce the primacy of the human species. Moreover, cultural and technological evolution are mutually dependent: all historical evidence indicates that, within the contingency of history as a human activity, cultures with more effective and efficient technology tend to outcompete others. Thus, the Bronze Age replaced the Neolithic (New Stone) Age, and was in turn replaced by the Iron Age, and so on until our current era, when globalization clearly reflects the ascendancy of capitalist, primarily Euro-American culture. In general, then, more advanced technology will, through differential increases in cultural fitness, dominate.

These observations do not mean that such dynamics are desirable. Rather, they respond to the "what is" question, not the "what should be" question. But trying to answer the latter without clearly perceiving the former can only lead to significant conceptual mistakes—and policies that just don't work.

Has this happened to some elements of environmentalism? Suppose, for example, that the species' will to power exhibited throughout history is foundational to being human—is environmentalism in opposing this simply making itself irrelevant? And if the confluence of that will to power and technological and cultural evolution has led to an anthropogenic world, can environmentalism evolve itself in response, rather than indulge in ideological denial? Further, "sustainability" tends not to be comfortable with technology, cultural conflict, or evolutionary dynamics . . . can it evolve as a viable construct, or is it condemned to be an increasingly marginalized cultural critique?

I do not pretend to know the answers. But asking these questions suggests that at the least environmental education is seriously deficient to the extent it does not embrace a study of history and the human condition—which it usually does not. Additionally, such an approach suggests that it might be wise to develop a new and dynamic environmentalism, one comfortable with technological and cultural evolution and humans as they appear to be . . . just in case history turns out to be a little better indicator of the future than we think.

Values, Goods, and the Environment
(June 2001)

The results of a *Newsweek* poll taken in mid-May deserve some thought on the part of environmentalists. They showed that, while a plurality did not think that the U.S. administration (the second President Bush) was "committed" to protecting the environment (47 percent versus only 39 percent that felt it was), a plurality also supported the administration's environmental performance (45 percent approval versus 41 percent disapproval). Moreover, 52 percent said that developing new sources of energy should be more important than protecting the environment.

These results can be taken in several different ways. For example, European deep greens will undoubtedly see them as confirming evidence of their belief in the immorality of American society in general. Conversely, industrial interests opposed to action in response to global climate change may well view them as indicating a proper prioritization of issues. Many, however, knowing that Americans strongly support environmental quality, will be puzzled. What should we make of this seemingly anomalous poll result, especially coming right after the well-publicized criticism of the American withdrawal from the Kyoto global climate change negotiation process?

The answer lies in the difference between values and goods. Values are the fundamental principles that form the cultural underpinnings of a society or culture. In the United States, for example, one would consider the rights enumerated in Bill of Rights as explicit values, underlain by the liberal principles of government expressed by such Enlightenment philosophers as John Locke. Goods, on the other hand, are recognized as important and valuable, but contingent upon context. Thus, education is clearly regarded in most societies as a desirable good, but economic and practical limits on how much is spent on education are universal.

Not everyone, even in the same society, will have the same hierarchy of values and goods. Moreover, the prioritization of goods and values within their categories will differ among individuals as well. Especially where individual commitment levels are high—and almost by definition commitment levels in NGOs are high, as that is what gave rise to them in the first place—goods may well be perceived to be as fundamental as values. In this particular instance, for example, environmentalists and associated policymakers, primarily in environmental regulatory agencies at the national and supranational levels, are usually self-selected and have much more of a tendency to regard environmental goods as fundamental values. Thus, many environmentalists, especially deep greens, elevate the dictum of Aldo Leopold—"A thing is right when it tends to preserve the integrity, stability and beauty of the biotic community. It is wrong otherwise."—to the level of a, even the, principal value. Because of the internal cohesiveness of the environmental community, the idea that the environment is not a value, but something more contingent—that is, a good—is not often raised, and is rejected if it is.

These poll results remind us, very simply, that for most people the environment is a good, and not a value. Thus, Americans will support strong environmental regulation, but when such regulation begins to conflict with perceived values, or even other goods perceived as more important (private property rights, the ability to support a family, economic stability, energy security), the environment will be prioritized. And for many it will not be a core value. (The same lesson can be drawn from the widespread recent European demonstrations by truckers and others against rising fuel costs: no government, not even in Germany where the Greens are part of the ruling coalition, made an environmental argument in favor of higher fuel prices.) The poll results support this reading as well: a majority of respondents supported both a strong energy policy *and* more energy conservation. To view these results as a rejection of environmental consciousness, then, is incorrect; rather, one detects a sophisticated effort on the part of the public to support a policy system that both reflects their values and, at the same time, recognizes the environment as an important good.

Why does this matter? It matters because a failure to understand the ways in which the public views environmental issues can lead to dysfunctional policy initiatives and, over the longer run, a loss of credibility for environmental management in general. An insistence that the public accept the environment as a critical value—indeed, in the view of some advocacy groups, the only value—when they clearly don't is setting the stage for a policy smashup that would be good for nobody: not the environmentalists, not society—and, importantly, not the environment.

A Government in Exile
(February 2002)

It is a commonplace of politics that a party that shines when functioning as the "loyal opposition" is frequently incapable of governing when it finally assumes power. The underlying reason, of course, is that functioning in opposition, when one is not responsible for programmatic execution and one's casual remarks are not likely to roil foreign relations and financial markets, requires different skills and mindsets than actually governing.

This simple observation has some significant implications for environmentalists, whether in NGOs or environmental organizations within firms. For modern environmentalism is largely the classic loyal opposition: it was born in opposition to late industrial society and had to fight hard against powerful economic and political forces during its formative years. While the results of that effort—cleaner air and water in developed countries, much higher levels of concern about environmental issues globally—are both apparent and desirable, the environmental movement is having significant adjustment problems even as it succeeds in some areas. It is a discourse that excels in attack and opposition but falters badly in governing (not to overgeneralize: notable exceptions in the United States include NGOs such as Resources for the Future and the World Resources Institute).

Consider as an example how many environmental groups, especially the deeper green ones, require an air of crisis for their continuation: admitting success threatens their very existence. If crises do not exist, they must be manufactured; risks that may be manageable at a relatively low cost to society must be made to appear far larger and more dangerous than they really are. This relatively common dynamic carries the danger of the Chicken Little syndrome: people may get so used to environmentalists claiming that the sky is falling that they will ignore all environmental problems, even the real ones, until it is too late. More subtly, this process skews the allocation of social resources, with impacts that are very real and undesirable but that can unfortunately be hidden. No one thinks about the poverty that could have been alleviated by funds that went to an unnecessary environmental regulatory program.

Another illustration of environmentalism's failure to constructively engage wih social change is the adamant opposition of many deeper green environmental groups to technological evolution. In the global warming negotiations, for example, deep green environmentalists continue to try to sweep virtually all technological solutions off the table. Geoengineering projects that could potentially mitigate the effects of enhanced CO_2 in the atmosphere are not even allowed on the agenda (the point is not that such schemes may not be problematic, which they indeed might be; the point is that even discussion of them has been stifled a priori). Biotechnology as an effective means of developing biological carbon sequestration systems (rapidly growing tree plantations, for example) is also completely unacceptable, both as a specific solution and as a general technology. Geophysical carbon sequestration, which would permit use of cheap coal reserves to help developing countries grow their economies, or iron fertilization of midoceanic planktonic communities, which could sequester some carbon—highly questionable. Nuclear power—unacceptable. The point is not that such options don't need a thorough vetting—of course they do. The point is that, for ideological reasons, they are not even considered. Even as new technological solutions are developed, they are rejected.

Why is this? In large part, the underlying reason appears to be that such technological fixes do not require people (especially in the rich West, and especially in the United States) to change their lifestyles sufficiently. In other words, as success has been achieved in many—not all—areas, the focus of environmentalism has shifted from reducing the pressure on the environment to social engineering. This also is a sign of an ideological loyal opposition, made more powerful by the aggregation in the environmental movement of various pieces of anticapitalist ideologies that have been forced undercover by the failure of Marxist-Leninist states. If one wants to oppose capitalism and globalization, deep green environmentalism is about the only good platform left.

But this posture is very dangerous for the environment. Many people in both developed and developing countries support environmental quality, but many of them, if they realize that they are being socially engineered to fit the deep green ideology of the moral life, will react strongly and negatively—and the baby of environmental protection and responsible management of the anthropogenic Earth may well be thrown out with the bathwater.

But I Want to Work on Environmental Stuff
(August 2001)

One of the horrible existential challenges of being a student is that, in most cases, one will have to at some point leave school and begin working, presumably in that area for which one has been training these many years. For those reading this column, the area of interest is likely to be environmental, usually expanded these days to include sustainability. Put bluntly, the relevant questions are likely to be: how do I do good, and what is the job market like? Recognizing that planning your career on the basis of a 750-word column is probably not a great idea, nonetheless here are some late-summer thoughts before you hit the books.

First, the good news. Although it may be hard in an entry-level position, there are plenty of opportunities to do great things: help your employer (be it private firm, government, or NGO), help the world, and feed yourself. But almost all of these opportunities are disguised, most have nothing to do with the environment as it is currently taught and thought about in most schools, a great many of them have yet to be invented, and almost any job that becomes worthwhile will require that you develop it yourself, from inside.

To begin with, traditional environmental jobs—that is, those based on current regulatory and policy structures, primarily cleanup and end-of-pipe emissions control—will be with us for a long time, especially in developing countries. They are necessary. But this field is not growing, offers few intellectual challenges, and will have little to do with solving the larger problems of the anthropogenic world (although it will improve health significantly in developing countries). So if you really want to help the environment in the broader sense—perturbed climatic and oceanic systems; anthropogenic carbon, nitrogen, sulfur, and hydrologic system changes; disruption of the biosphere—this is not the place for you.

The next step up is a position in the "sustainability industry." Superficially, at least, such jobs, which are frequently with avant-garde consulting firms, are broader in scope and offer more intellectual opportunities. But there are some cautions. The term "sustainability" has now grown to be so politically correct and at the same time has flown so far beyond mere ambiguity that there is no substantive content to a fair amount of this work. In too many cases, such work amounts to a somewhat patronizing, highly ingrown dialog within a small population that tend to regard themselves as the great and the good, and that spend a lot of time reinforcing each other's mental models. The result is a nouveau utopianism that has tenuous connections with the real world. Thus, for example, I recently participated in a sustainability workshop at a major U.S. business school where one of the products was the conclusion that firms should exist not for profit, but only to redistribute income—and that, by the way, money should be banned. Those of you with any historical background will recognize that this proposed policy closely tracks that of the Marxist-Leninist early days of the Soviet Union: they did ban money. And the economy collapsed. Moreover, you can imagine how the typical executive would greet such a proposal as a model for how his or her firm could become "sustainable." So be careful if you want to work in this area—you may want to work in a firm first, to get an idea of what companies are really like, before you jump in. It will help you keep your perspective. There are a few real opportunities, but—caveat emptor.

Right, then. So what to do? Back to first principles: the challenge of environmental (and related social) issues is precisely that they have grown to be so all-encompassing; they are not separable from the messy, multidisciplinary worlds of commerce, of ordinary life, of birth and death, of century-long natural cycles. So the kinds of things that contribute most significantly to social and environmental progress—telework options for employees, efficient network routing algorithms for air and ground transport systems, low energy and reduced water semiconductor manufacturing technologies—do not come from the environmental organizations, but from the core operating competencies, from the engineers, the business planners, the R&D shop. By all means remain committed to sustainability—but get expertise in international business, or chemical engineering, or finance. And then, when you get your nonenvironmental position—then you can start to change the world.

Green Technology: From Oxymoron to Null Set
(April 2002)

Last month's column raised the question of how the Internet and its post-modernist pastiche of time and place is liable to unpredictably change social perceptions and mental models of environmental issues, that have been firmly rooted in specific times and places. It thus hinted at another important insight that is complicating environmental analysis and policy formation: any technology complex enough to have real environmental benefits will be far too complicated to be understood—or in many cases even perceived—as a "green technology." While there are exceptions to this generalization, it is worth exploring some of its implications.

To begin with, this realization is yet another illustration of the poorly understood evolution of environmental concerns from "overhead" to "strategic." Environmental issues have gone from being treated only after the fact as remediation and compliance problems (as overhead), to being increasingly integrated into core activities of individuals, firms, and society (as strategic). In doing so, however, they inevitably lose their privileged status; they become only one dimension of the particular activity—and frequently not the most important one. For example, if I build a green accounting system for a firm, it does not mean that suddenly the environmental costs and benefits outweigh all the other data an accounting system generates. Rather, it means that information on environmental costs becomes another input to the decision process—and, depending on everything else the firm needs to consider, environmental issues may not be relevant or even addressed.

Turn now to the question of green technology. While such technologies are frequently defined quite broadly ("any technology that has environmental benefits"), in practice they tend to be end-of-pipe compliance and remediation technologies. Important, especially at a local level, for maintaining air and water quality? Sure. But meaningful in the context of an earth whose dynamics are increasingly dominated by a single species? Not really. And the concept hides an important dilemma: for a number of reasons, environmental professionals and environmentalists are poorly positioned to be able to understand, much less work with, technology as a discipline.

To understand this, consider some recent important "environmental" technologies. The problem: a high level of silver, an aquatic toxicant, in San Francisco Bay (and other aquatic environments). Research uncovers the source: photographic operations, including dentist and doctor offices. One possible solution, not very feasible but obvious given our current approach, is to try to regulate all relevant photographic activities and products. The real environmental solution? Digital photography. Or consider the environmental and social dimensions of e-commerce (an area most environmentalists have not even thought of). A critical technology for making business-to-consumer e-commerce more environmentally preferable is more efficient algorithms for solving the so-called traveling salesman problem—that is, how to visit a number of points in the most efficient manner. Is such a technology environmental? Sure—but it is also a lot more than that, and in fact the connection between efficient routing algorithms and the environment is intuitive neither for those who developed them nor environmentalists. Historically, the introduction of automobiles probably saved U.S. forests, which were being decimated by the need for land to grow oats to support an exploding horse population (the other major driver for deforestation around the turn of the century was the need for railroad ties, a demand that was overcome by the introduction of creosote to preserve wood, making that substance another ironic environmental technology). Telework may have significant environmental benefits—but it is being introduced because it enables significant improvements in business efficiency, while at the same time enhancing quality of life for employees and their families. Its eventual impacts could be far more profound, however: it is a technology of freedom for those who may be mobility challenged, including a rapidly expanding senior demographic—and it enables their ability to continue to participate in the economy as well.

These examples hint at several truths. First, any technology capable of producing significant environmental benefits likely has no obvious environmental dimensions, especially when viewed through increasingly archaic environmental mental models and orthodoxies. Second, any such technology will be sufficiently complex in its technical, economic, cultural, social, ethical, and environmental dimensions that interpretation through any single lens, especially environmentalism as currently practiced, will be problematic.

Implementing Strategic Environmentalism
(March 1999)

Environmentalism in corporations is an important driver of environmental quality, and thus it is important to consider what might impede such an evolution. This is not trivial, for the most powerful barriers to modern corporate environmentalism are cultural and institutional, and thus both subtle and difficult to manage. In fact, that is a major reason that, despite the apparent common sense of the concept, proactive and strategic environmental management has yet to take hold in most firms. But these barriers can, with patience and sophisticated management, be surmounted.

The most fundamental, and difficult, point for most individuals involved in such a transition to recognize is that strategic environmentalism is not an environmental organization function. In fact, to the extent that strategic environmentalism in any of its guises—design for environment (DFE), green accounting, life cycle assessment (LCA), industrial ecology, green design, or whatever—remains in the traditional environmental shop, it is probably an indication of failure.

Why? Because the essence of strategic environmentalism is the integration of appropriate environmental considerations into all relevant activities of the firm, from product and service design to operations, and virtually by definition that has not occurred if the environmental shop retains control and ownership. Consider, for example, DFE. In most organizations, if DFE guidelines come out of the firm's environmental organization, they will most likely be ignored, on the very reasonable grounds that the environmental group may know about environmental impacts, but is unqualified to speak of design. Similarly, if a green accounting methodology is rolled out of the environmental organization rather than the CFO's shop, it is also likely to be ignored, on the equally plausible grounds that the environmental group is not technically competent on accounting matters. As a final example, strategic business recommendations from an environmental shop, rather than from the firm's planning or marketing organizations, will most likely fall on deaf ears for the same reason.

This does not mean that the environmental group is not important. Indeed, initial championing of strategic environmentalism in most cases will only arise from that group, which is best positioned to both recognize and initially champion the concept. Occasionally a CEO will be taken by the concept, so a relatively straightforward top-down approach can be used (but do not underestimate the difficulty of converting middle managers, often the most resistant to any change, in part because they often bear the direct brunt of production quotas). But it is more often the case that a more subtle approach, based on a sophisticated understanding of internal firm dynamics, must be used.

This does not mean there aren't some obvious approaches that experience has validated in the past. For example, a useful tactic is to identify a target function, search for a champion in the appropriate organization, and feed the development of the appropriate tools through that champion. Thus, for example, when DFE was proposed for designing products in the (now divested) manufacturing arm of AT&T, an initial step was to understand how design standards and activities were undertaken in Bell Laboratories and to find a champion in Bell Laboratories who could introduce DFE into those activities. The end result was that DFE was fed to concurrent engineering teams and manufacturing engineers through Bell Laboratories—the institutional owner of the process—rather than from the AT&T environmental group, which would have been seen as overreaching and technically not competent in design (an accurate assessment). Similarly, in the network side of AT&T, when an initiative to explore more environmentally and economically efficient technologies for network power was begun, ownership of the project was vested in the network operations group, rather than in the AT&T environmental group. While environmental input to the decision-making process was obviously necessary, it was equally obvious that a decision on operating technologies, especially those involved in a high-reliability, complex technology system, should not lie with AT&T's environmental group, but with the company's network operation organizations.

This discussion makes two basic but critical points about implementing strategic environmentalism in the firm. The first is that, unlike traditional overhead corporate environmental work, strategic environmentalism is not about the environment as we all know it. Rather, it is about superior technology and business strategy. This is a difficult shift for many firms, both in concept and in practice. The second, also somewhat counterintuitive, is that implementation of strategic environmentalism does not require skill in environment matters as such. Rather, it requires the skills of a change agent and a sophisticated understanding of the specific firm, its culture, its markets, its strategies, and its knowledge base. The difficulties are many, which is why strategic environmentalism is so seldom seen in practice. The benefits—to the firm, to employees, to the environment, to society as a whole—are, however, extraordinary if it can be done, and make the game well worth the candle.

Brand Image
(August 2003)

Every year, Roper ASW does a survey on American attitudes and practices regarding the environment for the National Environmental Education and Training Foundation. This year's results? Overall, about 10 percent fewer Americans deliberately sought out articles or television programs, or otherwise tried to educate themselves, about environmental issues—the first drop in several years. Moreover, about 11 percent fewer were regularly trying to save electricity at home; 20 percent fewer were participating in any curbside recycling programs (only 36 percent were doing so overall); and 13 percent fewer were reading labels on pesticides in order to avoid environmental damage in use. Environmental issues seem to be holding less interest for the public—and it is being reflected in practice.

There are some obvious reasons for this. Recent economic performance has been problematic, and most people are more concerned about their economic well-being than all but immediate and severe local environmental challenges. Additionally, terrorism and the war against Iraq have clearly dominated public discourse, and a temporary eclipse of longer-range issues would be expected. But it is still a cautionary finding: it is the first time in years that data have indicated such a broad drop in concern about the environment. Moreover, it occurs at a time when many environmental groups are actively trying to encourage political engagement against what they perceive to be environmentally unsound governmental practices—and, with a few notable exceptions, are failing. This contrasts powerfully with the funding and public relations successes of environmental activists and NGOs during the Reagan administration, when environmental progress was similarly seen to be under threat.

It is at least possible that the public's shift away from environmental activism is not just a superficial and short-term phenomenon, to be reversed as the war news slips off the front pages. Indeed, there are a number of other straws in this particular wind. The market for environmental professionals is significantly weaker than it has historically been, a trend that began well before the United States and global economies ran aground. Environmental law and policy are becoming routinized backwaters, rather than the vibrant cutting-edge disciplines they once were. Despite desperate campaigning by activists, Americans continue to buy SUVs and light trucks in unprecedented numbers. On a more upbeat note, the UN Environment Programme has begun thinking about how to shift its presentation of environmental issues from a negative message ("These are all the things wrong with the environment and modern consumer practices") to a more positive approach ("Here's how to enhance your quality of life—and leave a better world for your children as well!").

This poses a difficult dilemma. Up until now, "the environment" has been a remarkably broad tent, including science, ideology, emotion, and belief; doctrines from Marxism, to antiglobalization, to antiwar, to corporate managerialism; and activists from the purely local to the international. It has been a traditional pattern that much activism has been sparked, and NGOs supported, by routine reference to crises and looming disasters, some more valid than others (compare the hysterical tone of *The Population Bomb* with the equally scary, but much more solid, *Silent Spring*). Despite this inherent ambiguity, the concept of "the environment" has remained broadly popular.

It is potentially a major problem if the environment is now losing its credibility and strength as a concept. After all, even if negativism, confusion of goals, and activism may have turned the public apathetic, or even hostile, to environmentalism, it does not mean that the underlying phenomena don't exist, and don't require responsible and ethical responses. It just means that public support for necessary measures will be smaller, less informed, and more fragile, and environmental campaigns much more subject to hijacking by extremists or those interested in other discourses (anticapitalism and antiglobalization campaigns are popular among environmentalists just now).

Equally important, if the environment loses its brand value—its ability to mobilize positive sentiment, affect behavior, identify responsible policies and designs—there are no easy replacements. We would still have an anthropogenic world—one where the dynamics of "natural" systems and cycles are increasingly determined by human demographics, technology, development, and consumption. But we would have lost a simple and beneficial framework by which to understand that world, a touchstone for conceptualizing it, a compelling brand of significant value. Thus, the supreme irony might be that environmentalism would have turned out to be one of the most subtle, and difficult to manage, human dangers to the environment itself.

Annotated Bibliography

1. W. Cronon, *Uncommon Ground: Rethinking the Human Place in Nature* (New York: W. W. Norton and Co., 1995). This is a unique and courageous book that challenges environmentalist orthodoxy and, mostly for that reason, has made Cronon somewhat of a target for naïve environmentalism. It is not too much to say, however, that if you have not read this book, you do not understand the environment or environmentalism—and that without such an understanding, you are all too likely to be part of the problem, rather than part of the solution, regardless of good intentions.

2. D. S. Landes, *The Unbound Prometheus: Technological Change and Industrial Development in Western Europe from 1750 to the Present* (Cambridge: Cambridge University Press, 2003) and P. J. Hugill, *World Trade since 1431: Geography, Technology, and Capitalism* (Baltimore: Johns Hopkins University Press, 1993). If you want to understand environmental issues, you need to understand the technological and economic context of global development over the past several hundred years. These books are good starting points; there are others as well. But the point is an important one: a strategic sense of the environment comes not through the environmental discourse, but through the exploration of other discourses, especially the technological with an environmental perspective. In other words, if you really want to understand the environment, do not read the usual limited literature; it will confirm your belief structure, but it will not educate you. And environmentalism will not tell you why the world is the way it is. But you need to understand that, and these books (along with those by David Harvey, J.R. McNeill, Marx, and the like) will help you in that effort.

3. C. L. Redman, *Human Impact on Ancient Environments* (Tucson: University of Arizona Press, 1999). Many moderns are ahistorical, without a strong sense of the powerful cultural and historical tides that have shaped the modern world. Particularly now that environmental policies and dialogs begin to extend from a time frame of months or years (as in cleanups and building sewage treatment plants) to centuries (as in climate change engineering and management), this is an increasingly dysfunctional gap in perspective and knowledge. Redman's book is a good reminder that the process of terraforming our planet began much longer ago than most of us realize, and is a subtle and profound one to boot.

4. S. Sassen, *Losing Control? Sovereignty in an Age of Globalization* (New York: Columbia University Press, 1996). The human rights discourse has paralleled the environmental discourse in many interesting ways, and both are in some ways on the same journey from overhead to strategic for society. This little book focuses primarily on human rights issues, but the obser-

vations regarding governance, sovereignty, and power structures are equally applicable to environmental issues. After you read it, think about the extent to which some versions of environmentalism imply a necessary loss of national sovereignty, and whether that is a good or bad thing.

5. B. Allenby, *Industrial Ecology: Policy Framework and Implementation* (Upper Saddle River, NJ: Prentice-Hall, 1999). I would do a better job on this book if I were to write it now, but it is still a respectable introduction to many of the issues that arise as environmentalism evolves from overhead to strategic.

CHAPTER 4

Alice in Wonderland: Environmental Management in the Firm

Environmental management at the level of the firm is not a new function; it has been a focus of activity since at least the industrialization of Europe. The Industrial Revolution began with an integrated advance in the production of cotton textiles, which in turn required a revolution in chemical manufacturing to provide detergents, bleaches, mordants, and other necessary inputs. Partially as a result of textile manufacturing demand, and partially because it was so useful elsewhere (for example, in soaps, glass manufacture, leather softening, and in production of gunpowder and alum), the demand for alkali—especially sodium alkali—boomed, to be met eventually by the Leblanc process (patented in 1791). This process involved adding sulfuric acid to salt to produce "salt-cake" (sodium sulfate), which was then burned with coal and limestone (calcium carbonate) to form "soda"—sodium carbonate. The process, however, resulted in substantial air emissions of hydrogen chloride—and so, in due course, the Alkali Act of 1863 was passed by the English Parliament, requiring manufacturers to absorb the acid in towers designed by William Gossage. Paralleling the regulatory responses were industrial and market responses: for example, Tennant in the late 1790s perfected his process for making bleaching powder by absorbing chlorine in slaked lime. Production of the powder rose from 239 tons in 1810, to 5,719 tons by 1850. It is well said that the history of the chemical industry is one of turning residuals into product, again and again—and managing the new waste streams until the residuals can be turned into inputs for new processes.

This example illustrates several reasons why the firm is of special concern in any effort to understand environmental issues and policies. First, it is firms that by and large produce the physical products and materials upon which economic activity (and modern life) is based. This cannot be done without transport systems, consumption of materials and energy, and emissions of byproducts (well, unless the zero-waste nanotechnology factory becomes more than a fantasy)—and thus necessarily implies environmental changes. Moreover, this production activity leads to many of the most observable environmental releases, from hazardous waste sites to air and water emissions. Accordingly, both activists and the public have grown to associate firms, and especially manufacturing activities, with environmental impacts. In fact, it can be argued that this preoccupation has over time become somewhat problematic to the extent that it has led environmental activists and professionals to focus almost entirely on manufacturing, an increasingly anachronistic and dysfunctional focus given that most developed economies are now service-based. (This depends, of course, on how one defines "service," which is itself a difficult question; most developed economies range from somewhere around 60 percent to 80 percent services in terms of employment and GDP).

Second, firms are a logical focus of environmental policy because they are the repository of most technology and of much of the applied science that leads to technology. They are thus a major leverage point for integrating environmental and social considerations into the technology design and diffusion process. As a corollary, environmental progress is increasingly a product of the nonenvironmental decisions and strategic technologies of firms, making usual environmental policies and practices less productive at the margin. This is especially true of some service firms, where advances in core technologies can leverage environmental efficiencies across the economy as a whole (examples might include teleworking, the substitution of electronic for paper billing, and "green retailing").

One indicator of the increasing importance of firms is provided by the history of industrial ecology, a field that developed primarily in the industrial sector (founders of the field came primarily from AT&T, General Motors, and other industrial organizations). Like many similar terms, "industrial ecology" has many definitions, but the most common emphasize the focus on systems, rather than individual artifacts, and the critical role of technology (this definition is from the leading textbook in the field by Tom Graedel and myself, both of us with AT&T at the time):

> Industrial ecology is the means by which humanity can deliberately and rationally approach and maintain a desirable carrying capacity, given continued economic, cultural and technological

evolution. The concept requires that an industrial system be viewed not in isolation from its surrounding systems, but in concert with them. It is a systems view in which one seeks to optimize the total materials cycle from virgin material, to finished material, to component, to product, to obsolete product, and to ultimate disposal. Factors to be optimized include resources, energy, and capital.[1]

One of the most interesting things about this definition is that it does not mention the environment at all. Rather, it focuses on a systems view that includes materials, energy, and financial flows and, by mentioning economic, cultural, and technological evolution, suggests an approach inherently inclusive of both social and environmental values.

While there are some earlier mentions of the concept of industrial ecology, the field really began in the late 1980s when a number of books, particularly from the American National Academy of Engineering (NAE), began taking the systemic approach to mitigating environmental effects of industrial systems. This led fairly rapidly to the development of specific methodologies and approaches, such as design for environment (DFE), and life cycle assessment (LCA), that could be used to improve environmental performance of materials and artifacts. Subsequent research at leading firms, and organizations such as the NAE and the Institute of Electrical and Electronic Engineers (IEEE), has continued to extend the pragmatic engineering dimensions of industrial ecology, while the formation of the International Society for Industrial Ecology (IS4IE) and the founding of the *Journal of Industrial Ecology* have done the same for the more theoretical side of the field. The initial focus of industrial ecology, as with some elements of environmentalism, tended to be on manufacturing activities and physical products, but the scope of the field has widened as it has matured to include services, and to extend beyond the scale of the individual firm.

The pivotal importance of firms (and their technology) for the environmental discourse has led to two different, somewhat exclusive, approaches by environmentalists. On the one hand, environmental ideology and culture strongly support an antagonistic, or at least highly skeptical, attitude regarding firms in general (see "Environment and the Culture Wars" and "Working for the Environment-Industrial Complex"). On the other hand, some environmentalists increasingly recognize that firms offer the potential to make substantial improvements in environmental performance even as their focus

---

[1] T. E. Graedel and B. R. Allenby, *Industrial Ecology*, 2nd ed. (Upper Saddle River, NJ: Prentice-Hall, 2003), 18.

remains economic efficiency and quality of life. For example, a technology like telework not only improves environmental performance, but enhances productivity and employee morale at the same time. This positive potential is often unrealized, in part because the environmentalists who might help recognize such opportunities are often reluctant to work with firms to help them come about. It is also the case, however, that meaningful environmental benefits arise from highly complex technological and institutional innovations that, at least superficially, have little to do with the environment—and are thus unrecognized by the relevant environmental communities, including that of the firm itself (see "The New Environmentalist").

Interestingly, the opposite is also the case (see "Managing Complexity")—managers desperately trying to understand an increasingly complex and difficult world are generally unaware of the insights and lessons that a knowledge of environmental issues, policies, and institutions might help provide them. Much of business these days—and indeed, of environmentalism—is attempting to understand how change occurs in the very complex, integrated systems that characterize the globalized economy and how, and to what extent, it can be guided (see "The Change Agent in the Gray Flannel Suit"). This is a huge question, of course, but an important one, for how and under what circumstances such systems, with huge institutional inertia, can be guided is a critical area requiring much more sophisticated study than it has so far received. After all, much of environmentalism, and of public policy generally, is an effort to change the behavior of these systems.

The general concept of corporate social responsibility (CSR) has been explored by many; depending on who one talks to, it includes corporate policies and practices on everything from labor and human rights, to environment, to corporate governance policies and transparency of financial structures. A more limited concept is the triple bottom line (TBL) which urges firms to measure not just economic, but also environmental and social performance (see "The Siren Call of the Triple Bottom Line"). Such an approach has a number of positives: for one thing, it is similar to what firms have usually done (despite rhetoric, most firms have traditionally had some social and environmental dimensions to their performance—how much depends on a number of variables such as size, culture within which the firm operates, legal and regulatory regimes, and the like). For another, it is relatively easy conceptually to understand what a TBL approach means. But measurement, especially of the social dimension, can be quite difficult and involve highly subjective value judgments. Thus, in practice implementing the TBL is a difficult goal for the firm, although the leading ones attempt to continue to move forward even under such circumstances, rather than simply succumbing to paralysis by analysis.

Working for the Environment-Industrial Complex
(October 1999)

A while ago, I was reading an article on pollution prevention written by an ex-EPA consultant, and was both amused and somewhat surprised to see "industrial ecology" identified as industry greenwash. My first response, of course, was dismissive: didn't the author realize that meaningful environmental progress could be achieved only through such systematic approaches as industrial ecology and its implementation through, for example, design for environment (DFE) and life cycle assessment (LCA) methodologies? Indeed, pollution prevention as usually interpreted by environmental regulators is a singularly limited concept, a relatively insignificant extension of end-of-pipe approaches, and it requires something like industrial ecology to energize it.

But my initial reaction was both unfair and superficial. The author was not really reacting to industrial ecology as it has been laid out in existing texts, or as it is being implemented in some firms today. Rather, the article implicitly made an important point about the nature of "the environment" itself: the very concept of "the environment," or closely related concepts such as "wilderness" and "nature," is constructed from underlying mental models that may differ significantly and that carry very different policy and governance implications.

Thus, "industrial ecology" does not enter the environmental discourse as an objective concept (although industrial ecology studies strive for objectivity and good science). Rather, at least some environmentalists will see it as a response to growing political pressure by powerful administrative and bureaucratic systems embodying a belief system based on scientific and technical rationality—as, in short, a defensive thrust based on a state/corporatist managerialism mental model. Seen in this light, industrial ecology carries several implications that to many in the environmental movement may be problematic: a powerful (and polluting) elite co-opting "real" environmentalism; establishment of a playing field (high technology and industrial systems), which implicitly degrades the knowledge base and operational characteristics of traditional environmental NGOs; and, more subtle but all the more powerful for that, a vision of a future "sustainable" world based on a high technology, urbanized society as opposed to an agrarian, localized world with large portions off-limits to people (of course, people have differing conceptions of the ideal end state, but these two tend to be the schematic poles that bracket approaches to the environment in developed countries).

Obviously, environmental issues continue to pose applied and tactical issues for the firm. These are the daily grist for the environmental professional's mill—assuring compliance with a myriad of regulatory requirements, training personnel in relevant health and safety issues, designing and supervising control technology systems, and the like. While the yearning to have these activities be considered "strategic" is clear, the truth is that such jobs, while critical, are primarily operational despite consistent efforts to position operational tools and methodologies as strategic in the misplaced belief that this is somehow superior (see "Is Environmental Management Really Strategic" and "Environmental Management Systems: A Tool Whose Time Has Passed?"). But competent management of this important function can neither hide, nor answer, the deeper question of how to think about the environmental dimensions of complex technological systems when they do not seem to have anything at all to do with the environment at first blush. This is a real challenge, and is increasingly relevant as the debate on biotechnology and, now, nanotechnology illustrates; such an approach is a major element of any sophisticated debate about sustainability generally. It raises serious questions about how environmentalism can, as it must, embrace a more strategic alignment with industrial and technological forces.

Environment and the Culture Wars
(July 1999)

To begin with the obvious: we all understand culture and have our own beliefs and practices; moreover, any good business school will help its students understand the need for multiculturalism in managing a firm. But this is, after all, an environmental column, not one on diversity in the firm—so why a concern about culture?

The answer is twofold. As I have observed in previous columns, firms are being challenged to meet not just economic targets, but environmental and social ones as well—the triple bottom line (TBL). This requires a certain sophistication about what such targets might look like, and how they are constructed by firms, stakeholders, and society. Additionally, any intelligent management of the TBL and stakeholder relationships requires that you know how stakeholders regard you, and, in turn, that you know what agendas and goals matter to the groups you are dealing with. This is not just a simple matter of asking them: culture is far more subtle than that. In most cases, individuals and institutions are so embedded in their cultural traditions that they are not able to express them objectively; rather, they have a worldview or mental model based on their traditions, which dominates their approach to issues. Failure to develop an understanding of these dynamics can hinder communication and collaboration, or can even fan unnecessary and unproductive conflict.

To begin with, to work for a firm is to become part of an institutional tradition that is referred to in much of the literature as "state and corporate managerialism." It is characterized by a reliance on scientific and technical rationality, powerful administrative and bureaucratic systems, and a faith in market mechanisms and economic growth. While many in firms see themselves as value neutral, the cultural context of the firm is anything but, especially in times of rapidly evolving market conditions (for example, globalization) and technological advances that raise significant ethical and moral issues for many people (for example, genetically modified organisms). Firms represent political, economic, and technological power; they also control critical knowledge. Stakeholders, such as environmental NGOs, are well aware of these power and knowledge disparities—and, of course, may not agree with the cultural values embedded in the concept of the firm itself.

But it is also important to understand that environmental values are similarly cultural at heart. For example, much environmentalism focuses on the concepts of "wilderness" or "nature" derived from powerful cultural con-

cepts such as the Edenic or frontier myths.[2] These myths have a num significant implications—for example, they can support a view of the ronment where environmental quality is inversely related to the prese humans. This mental model helps explain one of the major gaps in m environmentalism—the failure to deal with urban areas and urban despite the fact that most people live in or near such areas. It also e the conflict between some representatives of deep ecology and enviro tal justice: the former believe that the environmental justice mover simply an effort to piggyback a failed socialist agenda on an increasing cessful environmental movement, while the latter view traditional e mentalism as racist, colonialist, and completely out of touch with th environmental issues that most people face in their daily lives.

An understanding of these cultural dimensions can at least po manager to begin asking the right questions when dealing with stakeholders. For example, begin by asking how you are perceiv whether power disparities are so significant that they are liable to dis stakeholder process unless managed (you could, for example, pro external validation of claims based on technological expertise). And important from your perspective, what are the mental models of th with whom you are dealing? What is the concept of "nature" towar they hope to advance, and is it one that, for whatever reason, exclud or economic dimensions that you, as a representative of a (pres responsible firm must consider? Are they only interested in "enviror issues, or are they also concerned with social and racial equality, or considerations lying outside the traditional boundaries of enviro activism? Understanding the cultural dimensions of environmental not a luxury, but a necessary means of supporting the collaborat discourses that must occur if we are to move closer to an enviro sustainable world.

---

[2] "Myth" is here used in the nonpejorative sense of a deeply held belief syste useful, if controversial, deconstruction of the ideas of "nature," "environmer "wilderness," see W. Cronon, *Uncommon Ground: Rethinking the Human Pl Nature* (New York: W. W. Norton and Co., 1995); for a similar and instructi struction of technology see D. F. Noble, *The Religion of Technology* (New Yo A. Knopf, 1998).

It was important, therefore, not to take that article as just a naïve rejection of industrial ecology and its promise, but to understand it as a reflection of deeply conflicting worldviews that were all the more critical for being implicit and, to a large extent, even unconscious.

These two mental models—call them the managerialistic (perhaps characterizing an industrialist's approach to the environment) and the Edenic— are not the only common ones. Other generic mental models might include the authoritarian (environmental crises require centralized authoritarian institutions); the communal (with the caution that some communities can be extraordinarily violent toward minorities and outsiders); the ecosocialist (capitalistic exploitation of workers and commoditization of the world are the source of environmental degradation); the ecofeminist (male exploitation of nature and women derive from the same power drive and must be addressed concomitantly); and pluralistic liberalism (perpetual open collaboration involving diverse interests is the proper process to achieve environmental progress).[3] Obviously, these are simplifications, and in many cases individuals may well hold messy combinations of various models. But all of these can be made to sound appealing, especially if one accepts their implicit assumptions and value prioritizations. They also all raise some very difficult questions. For example, ecosocialism is somewhat tarnished by the abysmal environmental record of Eastern European communist governments prior to the collapse of the Soviet Union.

The obvious question for the manager blessed with the opportunity to negotiate these minefields is: which one of these mental models is "right"? The unfortunate truth is that we as a society are not ready to answer that question yet. This is not just because most people—environmental professionals, environmentalists, regulators, industry leaders—are naïve positivists and are therefore unwilling or unable for the most part to recognize their own mental models, much less to respect other parties' mental models. It also reflects a disturbing and almost complete ignorance about the implications of each model for the real world. What levels of human population, of biodiversity, of economic activity, would each mental model imply? What kind of governance structure? Who would win and who would lose (more precisely, what would the distributional effects of each model be)?

The important point, I think, is not the correctness of any particular

---

[3] Readers interested in exploring some of these issues further might want to read David Harvey's *Justice, Nature and the Geography of Difference* (Oxford: Blackwell Publishers, 1996).

model. Rather, it is the need to understand that differences among stakeholders in environmental disputes may arise not just from factual or economic disagreements, but from differences in fundamental worldviews—and that, at present, our current knowledge cannot privilege any particular one. A little sensitivity to how one's position and practices are understood by others can go a long way toward facilitating collaborations that are both necessary and plenty difficult as it is. Before one too readily criticizes others, one should recall the Socratic admonition and know thyself—and thy mental models.

The New Environmentalist
(March 2000)

Perhaps the most famous thought experiment in economics is Adam Smith's pin factory, which he used to explain the concept of division of labor. According to Smith, a skilled craftsman could make pins one at a time, and the pins would be very good indeed. On the other hand, a pin factory, where different groups of workers performed different tasks, could produce a great many pins much more inexpensively, leading to much greater economic efficiency and wealth. The power of economics in modern culture is such that most people, even if they have never heard of Smith's example, are well aware of the benefits of division of labor. What most of them are not aware of, however, is that the process continues, at a higher level and more vigorously, throughout modern economies. And this dynamic has significant implications for all those who consider themselves environmentalists.

A recent article in *The Economist* discussed a phenomenon known as "factories for hire," wherein companies contract out manufacturing to specialized firms—in other words, outsource manufacturing.[4] Although patterns differ, many electronics firms considered by the public to be manufacturers, such as IBM, Ericsson, and Cisco, increasingly do less manufacturing and concentrate more on branding and marketing. Other sectors, such as automotive, toys, and clothing, are following suit. In other words, where division of labor used to operate at the individual level, it now operates at the firm level: one firm increasingly focuses on systems integration and design, one on manufacturing, one on services based on the resulting platform, one on marketing, and so on. Like many modern phenomenon, this firm-scale division of labor is enabled by modern information infrastructure, especially company intranets and the Internet.

And why is this done? Fundamental economic forces. For example, complex modern manufacturing systems place increasing value on development of specialized knowledge; a firm that does nothing but high-tech electronics manufacture will be able to outcompete one that tries to be all things. But there are more subtle drivers as well. For example, the absolutely critical role of "time to market" in many sectors is well-recognized by business folk—Japanese firms are respected around the world for their ability to compete on cycle-time and time to market—but not by those with no industrial experience. (Thus, for example, during the U.S. Clean Air Act amendment debates

---

[4] "Have Factory, Will Travel," *The Economist*, February 12, 2000, 61.

over procedures for permit modification, electronics firms argued that permit approval processes that might well last more than half a year were impossible where entire product lines lived and died in shorter periods. Many environmentalists, on the other hand, regarded all such arguments as nothing more than attempts to circumvent permit restrictions.) Finally, there is the flexibility element: contract manufacturing can be rapidly shifted as technologies change and market leaders shift.

It is interesting that contract manufacturing is not driven to any great degree by the reason some environmentalists might give—that is, to enable the shifting of manufacturing to localities where environmental laws are substantively weaker. For these sectors, at least, market presence and the increased efficiencies of operating in an increasingly vertically diversified mode—traditional economic drivers—are far more important than any marginal benefits suggested by differing environmental regulatory regimes (an observation that in many cases undercuts claims by industry that environmental regulations in themselves cause firms to relocate and jobs to be lost). Rather, the power of the traditional response reflects the fact that, to an environmental professional, all problems look environmental—and thus, evolutions in business practices tend to be attributed by them, often without analysis, to (often questionable) environmental motives.

But this naïve response has associated costs. Most significantly, it blinds environmentalism to the really important questions raised by such shifts in industrial structure. What are the real world, triple bottom line (TBL) implications of the contract manufacturing trend? What are the new leverage points to introduce design for environment (DFE) and other technical design and engineering methodologies to enhance environmental preferability of the system, from materials choice, to artifact manufacture, to service offering? What are the implications for international governance systems? Who makes the design choices within which the meaningful environmental issues are embedded? Relying on superficial ideological responses in such a complex system can marginalize the environmental voice just when it is important that it be heard clearly.

And this leads to the interesting observation that the great environmental achievements of the twenty-first century will in all likelihood not be made by the ideologically committed activist. Rather, such advances will come from the engineers and the MBAs, trained to be sensitive to environmental and social dimensions of their activities and to implement that sensitivity in the unprecedented complexity of the economies and cultures of the twenty-first century.

Managing Complexity
(November 1998)

It used to be easy. When you got your own factory, it was your barony. With minor exceptions, prerevolutionary France was the mental model (you were, for example, subject to the king's—a.k.a. CEO's—displeasure should you fail to make your numbers, but beyond that, you ran the estate). If you were the boss in a commercial operation, there was none could question your writ, assuming you stayed within the rather loose boundaries of the law and custom. Moreover, the world was much simpler: technology stayed put rather than mutating wildly, and you pretty much knew who your competitors were and who your customers were. The scale of your operations was also considerably smaller in many cases, so that you actually knew what you did.

Well, those days are gone, and the business student today would be well advised not to dream of them returning. As author Peter Senge, among others, has pointed out, the world that the modern manager faces is far more complex, sufficiently so that the cognitive skills and capabilities of individuals are often incapable of rational decision making under the circumstances. (Here, I define the complexity of a system as the amount of information required to uniquely describe it; an increase in complexity is therefore an increase in the irreducible information content of a system). Hence, modern approaches to management include fostering the "learning organization," with "knowledge-based" operations that create value added primarily from the knowledge of individuals and groups within the firm, rather than from traditional inputs such as capital, labor, and natural resources. Rather than dictate, "stakeholder management" is the preferred operating mode, both internally and externally: from authoritarian to collaborative, communitarian systems. Technology, of course, has gone bananas in most sectors: in a sense, every firm is potentially RCA, happily manufacturing vacuum tubes while the tsunami of transistors gathers irresistible force just beyond the horizon.

Even the most fundamental institution of the modern world, the nation-state, is changing in subtle ways. The comfortable model that dominated until quite recently—the nation-state at the apex of power, with other economic, technological, and social institutions subordinate—is giving way to a still ambiguous balance among private firms (especially transnationals); NGOs (especially in environmental and human rights areas); local, regional and virtual communities; and national governments. Technologically, socially, economically—the complexity of the modern business environment is qualitatively different than that of only a few years ago.

There are few road maps and case studies concerning how to structure and manage institutions under such circumstances, so any additions to the literature are valuable. Let me suggest a source of applicable insight that is routinely ignored by most business curricula: the environmental perspective, but not in the usual sense. Rather than involving the establishment of command and control regulation, which is both traditional and assumes simplicity (or, in other words, that regulators and stakeholders understand the system well enough to manage it by centralized mandate), I mean an environmental stance that involves management of ecosystems and natural resources under highly contentious circumstances. Familiar examples might include management of the Florida Everglades to support agricultural, mining residential, commercial, and conservationist interests; attempts to preserve the Chesapeake Bay given rapid population, industrial, and agricultural growth in the watershed; and the efforts to control degradation of the Great Lakes, which involve a number of stakeholders in two nations. In such cases, virtually all dimensions of the management problem are complex—there are numerous, often adversarial, stakeholders with limited knowledge and perspectives; the underlying physical and biological systems are highly complex and nonlinear; temporal and spatial scales cover many dimensions, with the cycle time of some cycles (such as the policymaking process) significantly different than that of others (such as many of the natural cycles). Additionally, any solutions to be viable must address and integrate the critical elements of political, social, cultural, economic, biological, and physical systems, and do so while each of those systems is itself evolving, frequently in response to previous management decisions.

Difficult as this task is, sets of institutions have evolved in response, which are well worth studying as models of adaptive management.[5] In the case of the Columbia River basin, for example, where salmon, hydroelectric power generation, Native American rights and interests, commercial development, and population growth created a potentially explosive combination, legislation, negotiation, collaboration, scientific research, and institutional evolution (for example, the Northwest Power Planning Council) have begun to create possible sustainable resolutions.

It is not just that business interests are directly affected in many such sit-

---

[5] These case studies and others are discussed in L. H. Gunderson, C. S. Holling, S. S. Light, eds., *Barriers and Bridges to the Renewal of Ecosystems and Institutions* (New York: Columbia University Press, 1995), which serves as an excellent introduction to adaptive management of complex human/natural systems.

uations (suppose you were a manager in the Bonneville Power Administration, for example). Perhaps more importantly, such real-world case studies can help the business manager understand and successfully navigate—"control" is far too strong a term—the complexities that each of us finds ourself dealing with in our daily operations in today's world. The good business manager can no longer confine his or her sightline to a defined fiefdom or set of processes. The lessons of ecosystem management are not limited to environmentalists, but, in fact, belong in the boardrooms of every transnational.

The Change Agent in the Gray Flannel Suit
(May 2003)

Admit it. If you are reading this, at some point you probably wanted to save the world. You might still want to. But the fact that innumerable people before you have had the same desire, yet the world remains stubbornly unappreciative, gives some pause. Perhaps the way lies through ideologies? After all, they provide the teleologies, the methods, the vision, and emotional motivation and support for world saving. But ideologies are essentially intellectual bumper stickers, and there is a certain lurking suspicion that the real world may be a little more complicated than that.

Yes, ideologies are maladaptive in the real world for some fairly obvious reasons. First, ideologies are necessarily radical simplifications of reality; indeed, this is part of their appeal. Collapsing complexity into ideology may be mentally and emotionally satisfying but, if the essence of the problems at issue derives in large part from the complexity of the systems involved, ideology is almost definitionally dysfunctional. Additionally, the constituents of any ideology lie in the past, even if they purport to address the future (perhaps as prediction, as in Marxism). They thus become problematic in a period of rapid and discontinuous change. More subtly, ideologies operate by elevating the power of a particular idea over the real, messy, and contingent world. They thus cut off dialog and openness to new ideas and in practice can be profoundly antidemocratic and antirational.

Well, then, assuming that at least making the world a better place remains a valid goal, what is to be done? Perhaps the most important first step is to recognize and accept the complexity of the cultural, technological, economic, and natural systems within which we now find ourselves, rejecting a wistful but misplaced faith in ideological bumper stickers. One can then ask the critical pragmatic question: how does change occur in complex systems, and how do I guide such change?

This raises profound issues, among them the question of whether we have free will in today's complex, technologically coupled world. But determining whether free will exists, a question that has bedeviled theologians and philosophers for millennia, is not a fruitful goal for a short column. We can, however, ask more practical questions, such as: do complex systems such as firms, governments, and technological cycles have leverage points as they evolve, where introducing change is more probable? The short answer is yes, but you have to be sophisticated in understanding their dynamics to take advantage of them. For example, if you are an internal stakeholder (an employee, say), fundamentally changing an institution when it is in a successful phase is quite difficult. However, when that same institution is under significant pressure—an academic institution that may be slipping down the league tables, or a firm in deep economic trouble, for example—you have a much better chance of achieving fundamental change. But even then, the changes proposed must align with both the implicit and explicit culture and goals of the institution. Institutions, like people, can be quite good at saying one thing and doing another: an understanding of their real (usually implicit) culture enhances success. Enron, for example, had an outstanding public statement of corporate ethics; their internal culture was clearly not aligned with this statement. External stakeholders, however, may find leverage points in other ways (boycotting or publicly campaigning against a consumer-good firm, for example). You need to understand the systems within which you are operating, and your position in relation to other elements of it, to be effective.

More broadly, the discomfort that many academicians feel in the presence of a highly complex and contingent world—which is to say, one that is not captured neatly within the boundaries of a specific discipline—leads to a failure to educate most students these days about perhaps the most salient characteristic of the future they will deal with and be responsible for—its complexity. That our understanding of this complexity is in its infancy is no excuse for pretending it is not there, a posture that is becoming increasingly irresponsible, almost unethical. Rather, our ignorance should be a goad for greater effort, and the development of appropriate educational materials and approaches. No student—business, environmental professional, engineer, whatever—should be allowed to graduate without at least one course in complex systems. Only then will we arrive at the understanding and sophistication that can generate rational, responsible, and ethical change—and, perhaps, save the world.

The Siren Call of the Triple Bottom Line
(February 1999)

Most readers of this book are familiar with the notion of the triple bottom line (TBL), which holds that companies should try to optimize not just economic, but also environmental and social performance. This is a seductive idea and, as a guiding principle, is not bad. Unfortunately, as with many other ideas in the environmental policy field (for example, sustainable development), it is being trivialized.

Our powerful desire for simplicity drives us to believe that such ideas are in themselves solutions. Sometimes they are—but only in cases that were trivial to begin with. In a complex world, they can be guides, but they cannot render the complex simple. We do no one—the environment, social justice and human rights, and the economy—any favors by pretending otherwise.

There are instances where economic, environmental, and social interests align. We benefit, for example, by teleworking. If implemented properly, it reduces pollution and greenhouse gas emissions, improves peoples' quality of life and sense of community, and even helps traditional commuters, who face less congestion. Firms supporting teleworking get happier, more productive, more loyal employees (at least that is what the data show), and require less office space. Here, the TBL test is easy: all indicators point in the same direction.

But these cases are the exception. It is usually difficult to determine what is better for "the environment," which is, after all, not a monolithic thing but an integrated set of physical, chemical, and biological systems at all scales that are valued differently by different people. As for social issues, what is considered equitable and fair often is impossible to determine unambiguously. Whose interests are paramount? Traditionally, it is the nation-state that determines the interests of its citizens, but its decisions are increasingly being challenged by NGOs, or else benefit only the ruling elite. And when one group is upholding environmental goods as against significant economic interests of other groups, who should win?

Consider as a hypothetical example a campaign by an environmental NGO to protect an old-growth rain forest by demanding that targeted companies switch to nonwood fiber for the paper they consume and by boycotting wood and paper products from the region in question. The campaign will be enforced by publicity and direct action targeting firms that do not comply. From an economic perspective, the targeted firms, which only purchase paper but do not produce it, are relatively unaffected in the short term. In fact, the easiest path for the firms is to agree to the demands: no large firm today relishes being targeted by environmental activists, and it may cause damage to the brand.

From the environmental perspective, such a campaign is much more problematic: there is significant evidence that from a systems perspective the use of nonwood fiber in paper making is worse, not better, for the environment. In this light, consider what the response might be if it were the paper industry, not an NGO, that proposed a major shift in technology from one whose environmental impacts were well characterized to one that might be highly problematic.

The social dimension is yet more difficult to assess: boycotting an entire region that may depend on wood and paper products for its economic activity and employment may cause significant economic hardship and community and personal disruption. The social dimension is the most difficult to understand, primarily because very few objective metrics exist to identify optimal social solutions.

In sum, the TBL is an admirable template for assessing complex situations, but only if its limitations and conflicts are recognized and not glossed over. All of us—firms, NGOs, communities, and governments—should aspire to more rational, desirable behavior, and this concept can help. It cannot, however, make the complexities of the world simple if adopting such an approach is a substitute for responsibility and integrity. The obligation of every institution to perform in a socially, environmentally, and economically responsible manner cannot be reduced to superficial slogans.

Is Environmental Management Really Strategic?
(January 2003)

There is a powerful tendency among environmental managers, and the academic institutions that train them, to insist that environmental management is increasingly strategic. In fact, most environmental management is, and should remain, profoundly tactical. This does not mean that operational environmental management is unimportant: making sure that firms comply with applicable regulatory requirements is a constant effort, even when those requirements are relatively stable, as employees, processes and business conditions change. New technologies require new compliance and environmental management techniques. Nonetheless, environmental management is best characterized as an overhead function, relatively independent of the core, strategic missions of the organization.

This is not to denigrate the importance of the environmental function. Compliance with the law, especially in areas such as the environment where society has deliberately reduced the boundary between civil and criminal penalties, is not a trivial matter for any firm or manager. But it does suggest that most efforts to conflate environmental management with strategic issues for the firm are misdirected and arise from confusing "important" with "strategic." Environmental management is important. It is, in most cases, not strategic.

Take, for example, environmental management systems (EMSs). Are they important? Yes, in that they can help a firm perform its environmental (and frequently safety) functions in ways that are efficient, and, by being better organized, that are more likely to support compliance and avoid inadvertent violations. EMSs are an important operational tool, and any reasonably sized firm that does not have an EMS of some sort is being mismanaged. But that is not the same as saying that EMSs are strategic: the very fact that they focus on environmental performance demonstrates that they are not. It is almost oxymoronic: a truly strategic EMS would be strategic precisely because it would not focus on environmental issues at all. The idea of "strategic environmental management" may appeal to consultants, and to environmental managers who somehow think that a strategic management position is preferable to a critical operational one, but it is close to a null set in practice for most firms.

That does not mean that there aren't strategic issues that have environmental dimensions. For many sectors, from information technology, to biotechnology, to extractive industries, operations are increasingly challenged by a broadening range of stakeholders. Human rights and environmental issues that previously resided with political entities are increasingly in play as global governance structures shift in unpredictable ways. Examples might include current disagreements about genetic engineering or appropriate human rights standards for manufacturing in developing countries; where these would previously be resolved by the national governments in question, they are now debated by firms and NGOs and resolution may not involve the relevant governments at all. Such broad corporate social responsibility issues may well be critical to the firm, even leading to questions about its right to exist, at least among activists.

Another example is the transition of the firm from a facilities-based to a netcentric structure by such mechanisms as replacing reams of paper and centralized white-collar offices with network infrastructures based on virtual private networks (VPNs), and in which functionality is shifted to the Internet and e-business solutions. New work practices appropriate to a knowledge economy, rather than to a manufacturing economy, would include telework, virtual office packages, nonplace- and nontime-based operations, and knowledge performance assessment metrics. Clearly, this evolution has the potential for substantial environmental and social benefits if properly implemented—for example, significantly reducing paper consumption by the firm as the intranet replaces written documentation, and reducing unnecessary employee travel. Just as clearly, it is a challenge to the very definitional basis of the firm, and its implementation thus involves organizations from operational units, to human resources, to the information technology function, to corporate security. And it is strategic in every sense of the word.

These examples all have an environmental dimension, but are far more complex than that. Evaluating their impacts on the firm, employees, stakeholders, and cultures within which the firm is doing business requires integrating different kinds of substantive knowledge from many different sources. Environmental input may be relevant, but even then it will usually not be compliance and remediation knowledge, but broader environmental and stakeholder management expertise, that is important. For the student, the point is not that environmental knowledge and sensitivity are not important—it is that, if you want to participate in the strategic decision-making process of firms, do not train or job-search to end up in an environmental slot. "Important" is not "strategic."

Environmental Management Systems:
A Tool Whose Time Has Passed?
(May 1999)

Anyone who has any familiarity with current environmental practices and issues is familiar with the concept of environmental management systems (EMSs). These may be homegrown internal sets of processes for managing environmental and, usually, safety issues: virtually all firms of any size in developed countries will have such a basic system (and most will have fairly sophisticated ones). Such an EMS will usually be somewhat idiosyncratic, based on the specifics of the firm and its culture, and will be used primarily for internal management purposes. Recently, however, a lot of effort has been expended to create more generic, prescriptive environmental management systems, such as ISO 14001 and the European Union's ecomanagement and audit scheme (EMAS). Both of these gain in generality by reducing flexibility, especially in the case of EMAS, which reflects the formal EU process that gave it birth. Needless to say, both of these programs—as well as other variations on the theme—have generated substantial heated debate and consumed enormous effort. Are they worth it? Are these systems progressive, comprehensive answers to environmental issues, as they are usually portrayed?

EMSs as currently designed undoubtedly have some utility, especially in manufacturing sectors with a slow clock time, where layers of process do not inhibit firm performance or conflict with slower rates of technological evolution. This reflects the fact that modern environmentalism and its tools (for example, EMSs) have focused primarily on manufacturing and emissions from such activities and tend to be part of a culture of environment as "overhead," rather than encouraging integration of environmental considerations into all aspects of a firm's operations. But EMSs as generally proposed are relatively blind to services, which now constitute some 60 to 80 percent of developed countries' economies. EMAS, for example, is site specific: "the scheme is open to companies operating a site or sites where an industrial activity is performed" (Article 3). This works for a manufacturing firm, with a relatively few number of manufacturing sites with significant environmental implications. For a company like AT&T, however, with thousands of sites each one of which poses insignificant environmental risk in itself, EMAS is simply inapplicable. ISO is somewhat less rigid, providing for a range of approaches, but it is still a system better attuned to point, rather than nonpoint, industrial structure. More importantly, the value of service firms is the ability of services to enable discontinuous improvements in environmental efficiency across the economy as a whole, through innovations such as Internet publishing, teleworking, and efficient product and mail delivery algorithms. These enablers of environmental efficiency arise not from the traditional environmental function within the firm, which EMSs are a part of, but from the intelligent development and diffusion of core technologies, which EMSs are not designed, and are manifestly unable, to comprehend.

To sum up, then, the current proposals for generic EMSs would be potentially applicable to about 20 to 40 percent of economic activity, and, within that 20 to 40 percent, the vast majority of firms of significant size already have EMSs. EMSs fail to capture the vast majority of improvements in environmental efficiency; to the contrary, they continue the unfortunate positioning of the environment as overhead. It is hard to avoid the strong suspicion that, while in 1970 EMSs would have been an important policy initiative, they are redundant and passé in 1999. Depending on the country, they may provide interesting trade barriers, particularly for products from developing countries, but they will have a marginal, and probably not efficient, impact on environmental performance

So why the interest in generic EMSs? The principle drivers seem to be two: (1) a desire to show that voluntary, cooperative business/government initiatives work, and (2) a desire for transparency in corporate environmental performance so environmental impacts can be assessed. If these are the goals, EMSs would appear to be an inefficient means to meet them. There are other cooperative mechanisms, such as the Dutch covenant system, which would appear to offer more substantial performance improvements.[6] If the concern is improvement in environmental impacts, the best course is to develop environmental system monitoring capabilities, linked to research programs, that begin to tell us which economic activities are producing real environmental impacts, rather than simply gathering data and encouraging practices that may or may not be relevant. Company-specific EMSs are unquestionably useful, but the generic EMSs currently occupying the policy community are far more limited tools than most current commentary is ready to admit.

---

[6] For a critical comment on collaborative programs as a whole, see K. Harrison, "Talking with the Donkey: Cooperative Approaches to Environmental Protection," *Journal of Industrial Ecology* 2, no. 3: 51–72.

Annotated Bibliography

1. R. Socolow, C. Andrews, F. Berkhout, and V. Thomas, eds., *Industrial Ecology and Global Change* (Cambridge: Cambridge University Press, 1994); T. E. Graedel and B. R. Allenby, *Industrial Ecology*, 2nd edition (Upper Saddle River, NJ: Prentice-Hall, 2003); and R. U. Ayres and L. W. Ayres, eds., *A Handbook of Industrial Ecology* (Cheltenham, UK: Edward Elgar Publishing Limited, 2002). Industrial ecology is the study of industrial systems and their relationship to, and integration with, natural systems. Each of these books provides a different perspective on the field: the Socolow et al. volume looks at industry primarily from the viewpoint of the natural systems involved; the Graedel and Allenby book remains the leading engineering textbook in its field, and thus provides an accessible introduction; and the Ayres and Ayres volume is useful for reference and more-detailed coverage of specific topics. As usual, the edited volumes vary in quality across contributions. Those of a more scholarly bent who want to keep up with the field should consider subscribing to the *Journal of Industrial Ecology*, an excellent publication edited at Yale.

2. P. M. Senge, *The Fifth Discipline: The Art and Practice of the Learning Organization* (New York: Doubleday, 1990). Business and management literature is a phantasmagoric area, characterized by a strange and inexplicable tendency to leap from fad to fad, coupled with a magical ahistorical ability to forget that all the previous ones didn't work either and to label everything with single-digit numerical qualifiers ("Fifth Discipline," "Seven Habits," and the like). Nonetheless, this book makes an important point in a rather approachable way: the world within which business is trying to manage has become much, much more complex, and the human psyche is not well equipped to intuitively understand such complex structures, especially their dynamic behavior as they evolve over time. Besides, "the learning organization" has become a buzzword in its own right, and if you read this, you can say you read the source material. That's good filler for staff meetings.

3. T. L. Friedman, *The Lexus and the Olive Tree* (New York: Anchor Books, 2000) and M. Castells, *The Rise of the Network Society*, 2nd edition (Oxford: Blackwell Publishers, 2000). If you are going to understand industrial behavior, you need to understand what industry looks like—which means you need to understand the global economy. There are a lot of books about that subject, but among the best are Friedman's readable introduction to the global economy and some of its political and cultural implications, and Castells's more rigorous and in-depth exploration of the network structure that underlies it. Those who enjoy Castells's introduc-

tory volume might also want to read the other two books in the series: *The Power of Identity* and *End of Millennium.*

4. T. E. Graedel and B. R. Allenby, *Design for Environment* (Upper Saddle River, NJ: Prentice-Hall, 1996) and E. S. Rubin, *Introduction to Engineering and the Environment* (New York: McGraw-Hill, 2001). Two books that may be too technical for some, spanning the industrial technologist's approach. Design for environment (DFE) is a module of a design methodology called "design for X," where X is a desirable characteristic of a product (testability, manufacturability, and so on); it represents an effort to directly inject environmental considerations into the design process. Rubin's textbook is a thorough and more quantitative approach to engineering, with emphasis on environmental considerations. Neither is an "environmental engineering" text: as currently practiced, that discipline is fairly narrow, focused on remediation and emissions control technologies, and thus it tends to adhere to an environment as overhead worldview.

5. There are all kinds of books about how business resembles a garden, how to manage your company's debt load as if it were a grape vine, the relationship between cash flow and blood circulation, and the like. These tend to confuse anecdote for data, to be quite naïve about human institutions and how large firms operate, and in many cases they unintentionally (?) embody a gently Marxist, quasi-utopian Edenic teleology. To the extent they rely on a suggestive analogy between industrial and natural systems that can be explored for ideas, they are useful for the beginning student; to the extent they purport, like similar efforts over the centuries, to point to a future paradise similar to the golden age that we left behind (and which, in fact, never existed), they are relatively uninspired fiction. You can choose your own from among the many competing titles.

CHAPTER 5

Thoroughly Modern Marxist Utopianism: Sustainability

Much thinking is done through the use of cultural constructs, concepts that capture complex realities in simple ways that align nicely with a particular time, place, and culture. The most powerful cultural constructs simplify reality down to terms that embed within them entire discourses. Accordingly, when they are used, such terms necessarily impose a certain mindset and ideology on a discussion. Marxism was always good at this: to argue with a committed Marxist meant to use a language that was different—full of "Fordism" and "dialectics" and the like—which imposed a certain worldview on the entire debate. Similarly, contemporary American conservatives have a powerful and internally consistent worldview as well, and their language—"liberal," "family values"—has the same dominating effect.

So, too, with environmentalism. When environmentalists discuss "wetlands" or "rain forests," they impose a worldview on the discussion that limits potential discourse: to disagree with the environmentalist position in such an instance requires challenging the belief system and value structure built into the language of the discussion itself, a difficult and usually unsuccessful effort. Such a powerful effect has not, of course, gone unnoticed: Postmodernists have long noted what they call, somewhat theatrically, the "terroristic" effect of "dominant discourses" (a critique they usually turn on technology and science, and occasionally on capitalism and market-oriented economics, but seldom on their friends). But the truth remains: a good cultural construct creates its own reality.

Thus it is with what has unquestionably been one of the most successful cultural constructs coined in recent decades, "sustainability." This now-ubiquitous word was derived from the concept of "sustainable development," which was explicitly invented in its current form in 1987 by the World Commission on Environment and Development (the Brundtland Commission) in the book *Our Common Future*. There, sustainable development was defined as "development that meets the needs of the present without compromising the ability of future generations to meet their own needs."[1] The accompanying text indicates that this vision includes a high level of egalitarianism—that is, equality of outcome as opposed to equality of opportunity (or libertarianism)—that extends both within and between generations, as well as a high quality of life for all.

It is worth thinking a little about why the concept of sustainable development was thought necessary and about some of the reality that the term, now grown to cultural construct, brings along with it. To begin with, it is not a neutral term: at the least, "sustainability" implies some degree of wealth and income redistribution and thus coercion. It also carries with it a political philosophy (egalitarianism) that tends to be more comfortable for continental Europeans than for Americans—income and wealth redistribution is an idea that plays well in northern Europe but not in the U.S. Congress. The degree to which those urging sustainability fail to understand its normative content is both surprising and a significant reason why sustainable development initiatives tend to fare poorly in the United States (see "Sustainable Development: A Dangerous Myth").

Why was it necessary to invent the concept of sustainable development to begin with? A major reason was that there was an increasingly adversarial relationship between the economic development discourse and the environmental discourse, and a concomitant concern that if the two were not somehow aligned, one or the other would inappropriately dominate. Environmentalists feared that economic development would result in uncontrolled environmental degradation and population growth in developing countries, while development specialists and developing countries feared that unthinking application of environmental policies appropriate for developed countries would permanently consign developing countries to poverty, disease, and high mortality levels. Thus, creating the cultural construct of sustainable development was important because it implied that both discourses not only were compatible, but needed to be aligned if a sustainable world was to be achieved. And, indeed, it can be argued that the one fairly unambiguous achievement of the entire sus-

---

[1] World Commission on Environment and Development (Brundtland Commission), *Our Common Future* (Oxford: Oxford University Press, 1987), 43.

tainability discourse is that it has created a context within which productive, rather than conflictual, dialog between the two discourses is possible.

More subtly and implicitly, the sustainable development cultural construct and the sustainability dialog generally are also grounded in another discourse quite familiar to the Western tradition, Christian-Marxist utopianism. This is not necessarily bad, for much good has come from that tradition, such as respect for the individual and, more broadly, the modern human rights agenda. On the other hand, there is reason to be cautious about the unthinking application of sustainability as a guiding principal in day-to-day decision making, especially as the discourse is applied by powerful developed-country cultures to those that may have little opportunity to speak against it (see "Plato, More, Marx, and Sustainable Development" and "Temporal Imperialism"). Especially when sustainability remains somewhat ambiguous, the line between supporting human development and simply acting out one's personal ideological positions is not always clear. Thus, some in developing countries have complained that developed-country environmental NGOs are too eager to demand adherence to their environmental standards, and prioritization of environmental goals above others such as economic growth, under the guise of fostering sustainability, particularly in the use of resources.

Note, however, that a cultural construct, while it can be extraordinarily powerful, is still somewhat limited by brute reality: it can create new intellec tual topologies, but it cannot change the physics of the real world. Inventing sustainability does not guarantee that the "environment" and "development" goals and their underlying teleologies are or can be mutually compatible in the physical world. Consider, for example, precisely what it is that is to be "sustained." The vision of *Our Common Future* is an anthropocentric one, focused on the human as the priority, as one would anticipate from a construct that drew heavily from the economic development discourse. But there are many environmentalists who would reject this hierarchy, insisting that every form of life is equal (biocentrism). Moreover, there are deep green environmentalists and some Marxists who criticize sustainability as nothing more than a shell game designed to protect and continue current economic, institutional, and power structures (see "Privileging the Present"). Some of the Marxist critics (but certainly not the environmentalists) would argue that the implicit privileging of current forms of biological structure through, for example, antigenetic engineering activism, is equally questionable.

Some of this (surprisingly) reactionary flavor, and a focus on a relatively static utopian vision rather than on a dynamic concept, can be seen in the way many sustainability advocates view cities. In general, as William Cronon has illustrated in his classic *Uncommon Ground*, environmentalism tends to celebrate visions where people aren't, rather than where they are.[2] So it has always tended to view cities as centers of unsustainable resource and energy consumption (environmental-group calendars of the High Sierras or Last Great Wild Places are legion; no one publishes one on Sustainable Cityscapes). Although there are exceptions, many sustainability advocates have displayed the same propensity for unpeopled landscapes, although it is worth noting that some more recent groups, such as those involved with environmental justice, focus heavily on urban conditions. But cities, of course, are perhaps the most sustainable of human institutions; more than that, they are unmatched nodes of creativity, economic activity, technological evolution, and human freedom. To think about sustainability without thinking about cities—or worse yet, to classify cities as unsustainable in light of their ancient lineage and unparalleled evolutionary potential—is to shine a clear and brutal light on the pretenses of sustainability (see "Dynamic versus Static 'Sustainability'").

But below the angst of the ideological and class-struggle levels, a number of approaches have been developed that, although they do not claim the mantle of ensuring sustainability, at least enable improved social and environmental performance in existing practices and institutions. This is an important point to remember, for it helps avoid the danger of a currently undefined and unachievable best becoming the enemy of the good that can be done in the here and now. The most well-known of these approaches is the triple bottom line (TBL) methodology, which holds that institutions should attempt to optimize not just the economic, but also the environmental and social dimensions of their performance. The previous chapter discussed the applicability of the TBL to firms, but it is not immediately apparent why it should not apply to other institutions, such as NGOs, as well.

The TBL approach does indeed provide more easily understood conceptual framework than the more ambiguous one of sustainability within which to evaluate practices, initiatives, investments, and the like. But in firms it is a mistake to think the TBL easy to implement, for there is no guarantee that the three dimensions necessarily align, or, even if they do, that institutional and social cultural norms and expectations will permit progress. For that matter, it is often not clear that, even where opportunity exists, our institutional and cultural blinders will permit us to see it (see "The Twenty-Dollar Bill Fallacy"). Moreover, there are more fundamental issues that arise when the three suggested dimensions are considered together: for example, economic efficiency, environmental efficiency, and social efficiency (especially) are not only difficult

---

[2] See W. Cronon *Uncommon Ground: Rethinking the Human Place in Nature* (New York: W. W. Norton and Co., 1995).

to define, but require very different optimization techniques (see "The Efficiency of Inefficiency").

These thoughts on the TBL, and the broader agenda of sustainability, illustrate an important point. Despite the desire for clear direction and concrete advances, we are only beginning to see how to live in and manage an anthropogenic world. Ours is a period for experimentation, for making mistakes and learning from them, and for celebrating successes without the hubris of thinking that they are solutions. Ideology and rigidity of thinking are unlikely to be helpful in this process. An important first step, therefore, is recognizing that the tools and ideas we bring to the dialog about environment and sustainability are cultural constructs—that is, they seem absolutely true to us, but in fact they reflect a particular historical time, a particular place, a particular culture. Learning how to think about sustainability can provide a useful lens for trying to understand an increasingly complex world—but only if the context, assumptions, and inherently contingent nature of sustainability are understood as well. Such a posture not only enables further cultural evolution and experimentation, but also helps create useful citizens in an increasingly multicultural, increasingly human, world.

Sustainable Development: A Dangerous Myth
(June, 1999)

You don't want to knock myths, especially when they are popular. From the trivial to the profound, myths are the essence of culture and the framework upon which most of our reality is hung. So questioning them can be difficult and should only be done when there is a good reason. In this column, I want to talk about the myth of sustainable development and give two reasons why it has become a dangerous myth, one that can trip up those who fail to understand it.

The classic definition of "sustainable development" is that of the Brundtland Commission: "development that meets the needs of the present without compromising the ability of future generations to meet their own needs." As set forth in *Our Common Future*, it is a vision of a human-centered world of intergenerational, intragenerational, and sexual equality, with a high quality of life for all. Development is not eschewed, but occurs within the limits set by human and natural systems. An enticing view, all in all.

Several elements of this vision, however, are problematic in today's world. Note, for example, that the definition requires equality of outcome rather than equality of opportunity—that is, it is an egalitarian rather than a libertarian vision. In this, it not only reflects traditional northern European social democratic values, but the often-expressed desire of the developing world for income and wealth redistribution. Indeed, the roots of this vision go back farther: in *The Communist Manifesto* Marx and Engels wrote, "the free development of each will be the condition for the free development of all."[3] These historical antecedents are not pejorative—Marx, for instance, is a far more humane and insightful writer than Cold War stereotypes admit. But they do serve warning that the concept of sustainable development did not spring full-blown from the forehead of the Brundtland Commission, but rather is the offspring of a particular thread of political thought—which has generated considerable opposition as well. The term "sustainable development" is not neutral. Equally fundamental, perhaps, are realities such as the opposition of most of today's power structures to the kind of massive redistribution of resources contemplated and the opposition of significant religious and cultural elements to sexual equality. If these are prerequisites to sustainable development, history warns that we may never get there.

---

[3] K. Marx, *The Communist Manifesto* K. Marx and F. Engels, original published in English Edition, 1888, New York: Signet Classic, 1998, 76.

Indeed, many trends—increasing inequality of wealth between developed and many developing countries, particularly in Africa; increasing inequality within most developed countries—are moving in the other direction.

And that is one of the dangers of seduction by sustainable development: the term is being used so widely that it increasingly seems preordained. The danger that we are building a world that may be its dark twin—elites maintain power, sustainability occurs through loss of biodiversity and high mortality rates among the poor—is lost in the pleasant contemplation of . . . a myth. It is far easier, for example, to eulogize sustainable development in countless conferences and meetings, full of goodwill but bereft of content, than it is to address global poverty today. And yet the two should not be separate. And trends are going in the wrong direction.

The second danger is that sustainable development, by becoming so common and so comfortable, supports our inevitable human hubris. It lends an air of control, of knowledge, and protects us from having to recognize the profound depths of our ignorance when it comes to understanding sustainability. We need not struggle to leave the cave when the smiling, dancing shadows on the wall are so beguiling. Yet, we are ignorant in virtually every important dimension of sustainability, from the scientific, to the technological, to the social, to the ethical, to the economic, to the religious. Who knows what a sustainable nitrogen cycle would look like? Good question, not the faintest glimmer of an answer.

These dangers do not impugn the excellent work of the Brundtland Commission, nor should they be taken to imply that sustainable development is not a desirable vision. But they should be taken as warning: the concept is far more complex and culturally sensitive than most people realize. A manager, policy maker, or activist who chooses to use it, and related terms, needs to be aware of its connotations. Beyond that, integrity requires that those working in this area not succumb to the temptation to use sustainable development as a myth that replaces the need for true understanding and action. As a vision, it can inspire; as a myth, it is a dangerous soporific, yet another opium of the people.

Plato, More, Marx, and Sustainable Development
(May 2000)

Utopianism cuts a splendid, if somewhat raffish, figure in Western political thought. The ideals that typify much of this genre—egalitarianism (sometimes structured within classes), absolute devotion to an ideal, tranquility, development of human creativity and rationality—can be found early on in Plato's *Republic*, and, with appropriate Christian overtones, in works such as Francis Bacon's *New Atlantis* and Sir Thomas More's *Utopia*. Indeed, it is certainly arguable that Karl Marx's vision drew as much from the Enlightenment utopian tradition he inherited as from the economic and political philosophy of his time. Most people with only a passing acquaintance with Marx do not realize that much of his project was motivated by a desire to see people freed from what he perceived as cruel economic bondage—which, at that time, certainly characterized elements of industrialization—to be able to achieve their full potential.

Sustainable development as laid out by the Brundtland Commission, and subsequently adopted by many in various guises, is not self-consciously utopian or Marxist. Indeed, it speaks the language of science, of pragmatism, of human development. But at its core it carries with it an important link to the Western utopian tradition in its insistence upon egalitarianism: equality among and within generations. Egalitarianism is, quite simply, equality of outcome; the opposite philosophy, libertarianism, is equality of opportunity. In theory, the two are mutually exclusive; in practice, all societies achieve some sort of balance between them. All things equal, a level of egalitarianism is generally viewed as desirable if for no other reason than to avoid social strife. But, as the example of Marxism shows, egalitarianism as an ideology is both unrealistic and potentially dangerous.

This danger arises from a very simple dynamic, first raised by Aristotle against Plato's vision of the ideal city (which was, if you remember the *Republic*, communist in a number of ways): it is unnatural for people to give up that which they have obtained and feel is theirs. They must usually be forced to do so. And this therefore calls into being a coercive state or power. The more we demand people give up in the interests of egalitarianism, the more coercion we must be prepared to apply (the exceptions to this are societies that are largely homogeneous—Scandinavian countries, for example, or Japan). And even where this has been done by states with formal and ideological dedication to egalitarianism—Stalinist Russia or Maoist China, for example—there are still elites, often brutal ones. Egalitarianism in neither the material nor the power dimension prevails. The good ends desired by the egalitarian utopian are betrayed by the means necessary to achieve them in the real world.

Sustainable development, seen as an indication of the direction in which to evolve and a concept to encourage increased sophistication in environmental discourse, is a useful concept. Moreover, there is no question that the vast majority of the people working in the field today are concerned and committed individuals, seeking to build a better, more moral world. But there are those ideologues on the fringes who speak the language of coercion, of forcing consumers to consume less, of forcing the wealthy and powerful to give up their privilege. This is not just unrealistic, but is a dangerous undercurrent to the sustainable development discourse, one that may become more powerful as it becomes apparent that, certainly in any foreseeable future, egalitarianism will remain relative rather than absolute, absent complete social and economic collapse. In brief, if achieving sustainable development requires achieving a fully egalitarian society, then sustainable development will never be achieved. This is not to say that sustainable development, like other utopian concepts, does not have value as a vision toward which societies may aspire, only that those who believe it can actually be implemented may be somewhat overoptimistic. The utopian nature of the concept is, perhaps, the reason that the coercive and utopian elements of sustainable development are so little discussed—even recognized.

As with many of its utopian predecessors, a sustainable development world has much to admire—its idealism, its desire to see all peoples achieve their creative and personal potential, its egalitarianism. And the concept is an important goad to asking critical questions and envisioning alternative realities. But the threat of authoritarianism is a real one, and the experience of the late twentieth century with authoritarian, ideological states seeking utopias is uniformly negative. As Luke Skywalker's father, Darth Vader, told him: "You don't understand the power of the dark side of the Force." Beyond that, there is the danger that a useful utopian vision is being seriously misunderstood as a practical goal, with the result that much of the energy of environmentalism that could be important in helping shape the anthropogenic Earth is instead frittered away in trying to implement a fantasy. Here, as in all things human, it remains a constant challenge to keep the best from becoming the enemy of the good.

Temporal Imperialism
(January 2002)

The concept of "imperialism"—"the policy and practice of seeking to dominate the economic or political affairs of underdeveloped areas or weaker countries" (Webster's *New World Dictionary*)—is fairly familiar. A more recent variant, "cultural imperialism," where the domination is not military, political, or economic, but rather through forcing adoption of the dominant culture, is also fairly familiar: the French routinely charge the United States with cultural imperialism, and many in developing countries, especially in Asia, charge Western human rights and environmentalist NGOs with the same offense. Like most things cultural, such arguments are interminable, with common sense hounded by the politically correct, the nationalists, right- and left-wing ideologues, true believers of all stripes, and other hobgoblins of our enlightened age.

Accordingly, I will not discuss cultural imperialism. Rather, I want to talk about a related phenomenon, "temporal imperialism," that is as pervasive but much less explicit—as in, for example, the powerful effort on the part of present-day environmental elites to impose their views and ideologies on the future. My purpose is neither to support nor to attack temporal imperialism; rather, since the concept is so implicit and unconscious for most participants in environmental discourses, I want simply to note its existence and suggest that justifying such imperialism is an important and unfinished piece of ethical business. After all, the future is essentially powerless in the face of interest groups in the present, making temporal imperialism easy, but, concomitantly, a serious ethical challenge. Elites need to be particularly concerned about those, such as future generations, who have no voice at all (save what the elites, in their own self-interest, choose to impute to them).

Temporal imperialism is by no means rare. Just as historians have long noted the tendency of colonists, whether Polynesian or European, to carry their familiar landscapes with them through space, it appears to be a human desire to carry one's familiar cultural landscape through time. Thus, the desire of a powerful elite for stability through the ages has occasionally been quite successful—notable examples include ancient Egypt, China from 1400 onwards, and Spain from the 1600s on. But the cost of such stifling stability has often been the eclipse of the static culture by other, more dynamic, ones; freezing technological and cultural evolution reduces the fitness of a culture in competition with others. Moreover, most humans do not want to forego the benefits of technological and cultural evolution, which throughout history have tended to include increased freedom (from drudgery, from disease and poverty, from authoritarian community and cultural patterns).

And what are some current examples of temporal imperialism in the environmental area? One might begin with the powerful opposition to GMOs on the part of many deep greens: while scientific questions certainly exist and should be researched, there is little question that the opposition is in many cases absolute, nonnegotiable, and virtually theological. The precautionary principle, as commonly phrased, is another example: consider the language in the UN General Assembly's Resolution on the World Charter for Nature, which states that "where potential adverse effects are not fully understood, the activities should not proceed."[4] Since "potential adverse effects" are never fully understood, this language if taken literally would freeze technological, economic, and cultural evolution immediately—a signal victory for temporal imperialism. The climate change negotiations seek to stabilize climate within certain bounds, which has the effect of reducing or eliminating what has been in the past a significant driver of biological evolution. And, ironically, the concept of sustainable development might be interpreted by a good Marxist as an effort by existing elites to ensure their continued domination, and to impose a current cultural construct ("this is the world we want you to have") on the future.

Temporal imperialism is unavoidable (indeed, it is inherent in the very concept of culture) and in many ways may be desirable. What is of more concern, perhaps, is the failure of institutions and individuals in highly ideological environmental debates to perceive, much less understand, that what they seek to do to the future may reflect their own biases and predispositions, not what those living in those future societies may in fact desire. The remedy for unethical constraints on future generations is to optimize their choices, not impose ours, either implicitly or explicitly—and to recognize what we are in fact doing, for only then do we also recognize and begin to accept moral responsibility for our actions.

---

[4] UN General Assembly, Resolution on the World Charter for Nature, Resolution 37/7, 1982, Section II, para. ii (b).

Privileging the Present
(August 2000)

Many readers of this column no doubt remember fondly Dr. Pangloss, the eternal and foolish optimist of Voltaire's *Candide* who, like Leibnitz, concluded that this must be the best of all possible worlds. In general, one's personal experiences tend to argue that it probably isn't, at least all of the time and on an individual rather than collective basis. Accordingly, the Panglossian approach tends to be viewed with amusement and regarded as, at best, somewhat superficial. Ironic, then, that a mirror image of this frothy philosophic fantasy has become an implicit ideology of much environmentalism.

In particular, it is seldom recognized how powerful and unquestioned the drive to "privilege the present" is. Consider, for example, what the Kyoto process is really about: stabilizing the climate. If successful, this removes an important source of environmental variability that has encouraged the evolution of life on this planet from its beginning. In short, it dramatically affects future evolutionary pathways and, in doing so, strongly privileges not just present genetic structures, but present economic, social, and cultural structures as well. By what right do we do this? Or, perhaps more accurately, do we even recognize what we are doing?

Alternatively, consider the opposition to GMOs. Leaving aside questions of whether developed-country elites should deny such technologies to developing countries, it is clear that such a stance privileges present genetic systems—most of which have, in one way or another, already been altered by human activity. Or take an even more difficult issue: genetic engineering of human germ cells. Blanket opposition to this project amounts to absolute privileging of existing human genetic structures—and, concomitantly, existing ethical and theological systems. And of course the precautionary principle as formulated by some environmental activists—that no technology or practice be introduced unless it can be proved it does no harm—is another example of absolutely privileging the present. Note that in all these cases we cannot stop the underlying natural and human systems from evolving; human population and economic growth, if nothing else, will ensure that our species increasingly dominates many of the planet's natural cycles. What we can do is reduce the option space—the range of future potential choices—and ensure alternatives are not explored. This may or may not be desirable; we don't know until we at least ask the right questions.

The point is not that good reasons for privileging the present cannot be developed. For example, in many cases we are dealing with very complex systems, and the base state of knowledge concerning their structure and dynamics is frequently poor. Particularly where a proposed technology or intervention is significant and irreversible, therefore, a strong element of caution is highly desirable. But these are rational and operational considerations, capable of resolution through research and knowledge accumulation—they are not where the ideology of privileging the present comes from. The powerful emotional content of the dialogs around global climate change, GMOs, and the strong formulation of the precautionary principle tell us that something else is going on.

In part, it is a reaction against the constantly accelerating pace of change that characterizes the modern world, where everything from technology, to global governance systems, to existing institutions are in perpetual flux. Driving toward a static world where as many evolutionary alternatives as possible are eliminated is a natural response. Frequently these aspirations are described as returns to a Panglossian golden age: an age of morality, of stability, of prosperity, of defined positions in society, of biodiversity and Edenic pastoralism. The other side of this dynamic, however, is more problematic: an authoritarian and imperialistic imposition of developed-country ideology on the world as a whole. Privileging the present supports existing power structures, relationships, and elites. Granted, this is probably not the conscious desire of those involved, but it is still a predictable result.

And that, perhaps, illustrates the fundamental problem with privileging the present. It has significant effects, yet is seldom even recognized as a relevant dynamic, even though when and whether to privilege the present is an important ethical and, indeed, metaphysical and religious question. It should be an explicit, not implicit, part of discourses such as the Kyoto negotiations or adoption of the precautionary principle where it obviously pertains. We have, however, yet to take even the first step: to recognize that we are, in fact, privileging the present, and to explore and define, and then accept responsibility for, the implications of that position.

Dynamic versus Static "Sustainability"
(November 2002)

Although any attempt to define the term "sustainability" has fallen victim to ambiguity and co-option, that does not mean that playing with the idea is intellectually sterile. Indeed, it can be enlightening, as I found anew in a recent workshop in Los Angeles dedicated to the idea of "sustainable cities." In particular, two points became apparent: first, that any effort to understand sustainability in any guise without understanding cities is futile; and second, that cities demand a much more sophisticated understanding of sustainability than we have yet managed.

To begin with, it is frequently forgotten that cities—like them or not—are perhaps the defining icon of human evolution and civilization. Moreover, while we may be *Homo sapiens* or *Homo faber* (Marx), we are also Homo urbans: somewhere around half of all people live in urbanized areas, and that number is growing, especially in developing countries. Thinking about any kind of development—sustainable, unsustainable, high modernist, puce, or whatever—is simply an exercise in fantasy if urban systems are not part of it. Thus it is no surprise that there is a large literature here, with sweeps of conceptual thought based in virtually every philosophic school, from the "garden city" ideas of Ebenezer Howard, to the Platonic (and unworkable) geometric cities of Le Corbusier, to the high-modernist hubris of Robert Moses, to the counterresponses of Jane Jacobs and Charles Jencks, to Michael Dear's city as postmodern construct (the latter is particularly interesting: is Los Angeles, for example, a city about which films are made, or a concept from film and literature that happens to exist for a while in a built format?). And indeed Marx, and much more recently historians like William Cronon, have written of the complex interrelationships between cities, their hinterlands, capitalist economic structures, and the natural and human world. But there is very little literature or research looking at the industrial ecology of cities, or the intersection between the rich literature on cities and the much newer, and far less rigorous, literature on sustainability.

There are a number of reasons for this gap, ranging from the antagonism to cities in some traditional environmentalist thinking, to the simple fact that the sustainability discourse is clearly in its infancy. But an important element that may be difficult to overcome is that sustainability to many connotes a static achievement of perfection, while cities are preeminently complex structures—and thus can be understood only in dynamic and evolutionary terms. Show me a snapshot of a city, and it will inevitably be unsustainable; show me the grand scale of history, and human cities are the history of the species itself. Cities are sustainable only over time and within the dynamics of complex evolving systems, with all the contingency and reflexivity that characterize human history. Simplify them—turn them into the New Atlantis of Bacon, Utopia, the New Jerusalem of Europeans landing in the New World, or Le Corbusier's rational, geometric structures—and they are caricatures, stripped of reality.

There are, of course, elements of sustainability that imply a dynamic approach—even the term "sustainable development" implies some sort of continued development. Yet, sustainability for many people is a profoundly static teleology, expressed in such forms as the precautionary principle, opposition to technology, and a profound tendency to "privilege the present"— that is, to preserve the status quo in biological structures as well as human cultures. This static concept of sustainability, applied to urban systems, inevitably finds them deeply flawed. Cities are too messy, unpredictable, complex, and rapidly changing; they encourage commerce, economic growth, and cultural evolution; they tend to be dirty and require constant attention; they draw resources from the hinterlands and contribute nothing but air and water pollution back. They are living things, not utopian.

There is certainly no question that great urban systems are challenging: they challenge our governance capabilities, our understanding, our engineering capabilities. But they are, for better or worse, also the essence of our evolution, and to think that they will not continue to evolve in their own bumptious, unpredictable way is a crippling fantasy. Thus, they demand of us a new and more sophisticated concept of sustainability, one that is couched not in the tired, pessimistic, and defeated mindset that is so characteristic of our environmental dialogs today, but one that is dynamic and forward looking, that understands that the opposite of evolution is not sustainability but death.

The Twenty-Dollar Bill Fallacy
(January 1999)

All of us are neoclassicists. Whether we claim to be Marxists, structuralists, antieconomics generally, or maybe Taoists . . . we are all neoclassicists. In particular, we are so imbued with the myth of the market that we believe it: we believe in economic efficiency, in rational allocation through market mechanisms, in rational behavior in competitive firms. In short, we believe there are no twenty dollar bills on the sidewalk, because they are all picked up by the proverbial rational economic man, *Homo economicus.*

Maybe they have been, but probably not. Consider a case study of the triple bottom line (TBL), the effort to optimize not just economic performance, but also social and environmental performance as well. It is a nice concept, although it is not nearly as easy as superficial treatments would have you believe. But at least when all three dimensions line up, it provides a comprehensive framework in which industrial initiatives can be judged. Think of teleworking in this light.

Take the social benefits first. Teleworking empowers employees, so it directly enhances quality of life. It gives them time and money back. By enabling people to work in their neighborhoods, it enhances a sense of community and security. By reducing peak congestion on roadways, it enhances the immediate quality of life for those that must commute. Moreover, because infrastructure capacity decisions are driven by traffic peaks, it reduces the need for communities to build as many roads, saving open land and tax dollars. From the national viewpoint, productivity is maintained with less energy consumed, which helps everything from the balance of payments to national energy security. From an environmental viewpoint, teleworking has a number of benefits as well. Obviously, direct emissions from the commute that would otherwise have occurred are avoided, thereby reducing contributions to both global climate change and local smog. In addition, because those who do not telework spend less time in traffic jams with their motors idling, their emissions are reduced as well. And because teleworking firms generally use less office space per employee, the need for unnecessary construction, and thus unnecessary material use and environmental impact, is reduced. From a financial point of view, firms with teleworkers benefit because they are able to hold on to top employees, especially younger ones, who value the empowerment that teleworking provides. They also need less office space, which saves money. Moreover, teleworking helps them demonstrate their social responsibility to their customers and other stakeholders. Finally, the firm that provides the platforms and services underlying telework benefits directly from increased sales.

Leave aside for a moment the problem that it is at present difficult to quantify (or even verify in some instances) these potential benefits because of data uncertainty and peoples' adjustment over time to a teleworking job. (Will they make more shorter trips? Will the average distance between office and home increase over time?) A more fundamental problem is that this apparent home run—economic, environmental, and social signals all positive—has been trying to happen for years, and will undoubtedly struggle for many more. Why?

The answer is that institutions and managers that have not learned to see the twenty dollar bill will not pick it up. To begin with, the teleworking concept cuts against one of the most pervasive environmental attitudes: if it has to do with the environment, it cannot be positive. Thus, for example, environmental regulators who might have usefully pushed telework as an alternative to, say, MTBE in gasoline (which appears set to cost the nation billions in cleanup costs), have no such program. No one involved in the Kyoto negotiations on global climate change has raised teleworking as an example of a win-win mechanism to reduce carbon dioxide emissions. It does not fit the standard environmental conceptions.

Business is no less bound by culture and institutional perceptual boundaries. Managers who might increase the productivity of employees, and perhaps even retain valuable ones that leave, still believe that physical presence is a meaningful surrogate for productivity (amusing to see this relic of factory-floor management in today's knowledge-oriented firms). In a real sense, they cannot "see" the alternative. Even firms that might benefit frequently fail to see the opportunities. For example, the 1990 amendments to the Clean Air Act contained provisions that required commute restrictions in areas with bad air quality. Although this was essentially a potentially lucrative market created by government fiat and handed to firms whose services or products supported teleworking, they failed to take advantage of it. Why? Because their marketing organizations had never seen a market whose relevant dimensions were environmental and so, when it was before them, they also literally could not see it. Not one of them picked up that twenty dollar bill.

The lesson is simple and applies far beyond traditional environmental management. Any good business person, wherever they are, needs to be alert to the cultural and, especially, the perceptual limitations of their organization and relevant stakeholders. Successful firms and individuals usually focus over time on only those elements of the environment within which they operate, which experience shows to have been important. In periods of rapid change, this heretofore adaptive mechanism becomes self-destructive. The future is not built by the blind, and it can quickly crash down upon those who fail to see.

The Efficiency of Inefficiency
(July 2003)

"Efficiency" is one of those terms that everyone believes they understand but find surprisingly difficult to define precisely. There are a number of dictionary definitions, of which the most instrumental is "energy expended per unit work." Using a similar approach, "economic efficiency," for example, is technically the point at which the value of the marginal product equals its price, or, more generally, where the benefits from using an additional unit of resource equals the cost needed to acquire it. But the connotations of economic efficiency are broader. Many use the concept, for example, to justify free-market structures, favored policies, and global capitalism generally. Concomitantly, for many environmentalists the concept is pejorative, meaning globalized greed and industrial neglect of environmental values. An economist might respond that economic efficiency is intended to be a statement about how to obtain the most benefit from a resource, not a statement about cultural values and morality generally. And environmentalists would answer that it is used that way regardless. And so it goes.

In general, it is the connotations of economic efficiency in the real world that generate such heat: economic efficiency as a concept at least has an accepted and quantifiable definition. This is not the case with the question of the efficiency of the two other legs of the triple bottom line (TBL), environmental and social. If we cannot define these components in terms of efficiency, it very likely reduces considerably our ability to understand and implement a TBL approach. Yet here the arguments about economic efficiency begin to seem positively trivial.

To begin with, there is no accepted definition of "environmental efficiency" at the conceptual level. For one thing, much environmental doctrine rejects the idea that there can be any tradeoffs of environmental goods, or at least of those regarded as sacred or beyond price (for example, endangered species). In such cases, the marginal cost/benefit approach used in economics cannot be used to define an environmental efficiency. In practice, of course, life is a marginal game, so one can come up with heuristics along the lines of "consume as little material, toxics, and energy to produce the desired product or service, all else equal." Such approaches are valuable because they imply a marginal analysis ("try to produce a given quality of life only up to the point where the value of the environmental good used for the last unit is equal to the price of that unit")—but without directly requiring such analysis, and thus not challenging environmental orthodoxy. More directly, a number of increasingly popular policies, such as emissions trading systems and the like, operate by incorporating natural systems into economic systems, thus making them susceptible to marginal analysis and management through economic efficiency. But such approaches remain controversial, not least in the environmental community.

However, the complexities of "environmental efficiency" pale next to those of "social efficiency." Here also many of the conceptual and policy structures enshrine absolutes: "human rights" are generally justified by reference to "natural" law (or religion), not as a product of rational marginal analysis. But interestingly enough, it is perhaps the essence of social institutions, including governments, that they exist primarily to distribute costs and benefits that differentially affect their members in ways that, satisfying no one, at least achieve a kind of real-world balance. In this, social efficiency is perhaps unique, for it must be environmentally and economically inefficient to function. That is, it must reject the instrumentality of pure economic efficiency, while appreciating its importance for quality of life; and it must reject the ideological rigidity of environmentalism, while appreciating the importance of natural systems. And it must do so because it operates at a level of complexity above efficiency and environmental ideology, a level where, somehow, mutually exclusive positions and passionately held and conflicting beliefs must be integrated. Thus, for example, the U.S. government's proposed Everglades restoration project in Florida satisfied no environmental or economic criteria of efficiency. But from a social perspective, it achieved an integration that no single approach or discourse could have achieved, and thus maintained social cohesion—it was socially efficient in a fundamental sense.

The real lesson, however, may be more profound. Economic and environmental efficiency can be understood for humans only in the context of social efficiency—it is not a triple but a double bottom line, environmental and economic, embedded in a much more important social and cultural context. In other words, only through a process of efficient social construction can the meanings of economic or environmental efficiency be created, and validated. It is perhaps doubtful that the environmentalists who developed the TBL in response to the evolving idea of sustainability fully realized the degree to which environmentalism would inevitably become subsumed in the social.

Annotated Bibliography

1. Le Corbusier, *The City of Tomorrow and Its Planning*, translated from the 8th French edition by F. Etchells (1929; Mineola, NY: Dover Publications, 1987) and J. Jacobs, *The Death and Life of Great American Cities* (1961; New York: Vintage Books, 1992). Reading Le Corbusier and Jacobs together is a rare treat. It contrasts the high modernism and (some would say) hubris of Le Corbusier, who yearned for cities of geometric purity, with the almost postmodern approach of Jacobs, who understood and celebrated the informality, evolutionary qualities, and pastiche of viable modern cities—classic Plato versus Aristotle. And, although there is little question that Jacobs was the more accurate in understanding the messy world of urban life and the way cities actually evolved (and thus how planners could best contribute to the process), as with Plato there is an enchanting air of otherworldliness about Le Corbusier's proposals. It may seem to some that the latter's vision of a city—planned, orderly, controllable—is the more sustainable; that it is less realistic than Jacob's city, and has been impossible to implement in the real world, may therefore be a cogent comment on sustainability.

2. P. Hall, *Cities in Civilization* (London: Orion Publishing Group, 1998). Hall's book is without question a classic that anyone wanting to understand cities and their contribution to human evolution should read. It studies and indeed celebrates that which has been most characteristic of cities—their contributions to cultural and social evolution—and in doing so raises serious questions about any worldview that purports to understand the anthropogenic Earth without understanding cities. But it also rigorously illustrates troubling themes (raised elsewhere as well): for example, that history indicates that cultural progress occurs only when there is a significant degree of conflict (albeit one that does not slide into chaos and anarchy). Societies that successfully opt for stability, or have it imposed on them by a dominant institution, seldom prosper, and over time are often the fodder for the expansion of others. Those cities that are confused mixing bowls of various strands of humanity and thought from around the world—the Jacobs cities, one might say—are where human vision is created and reified.

3. F. Bacon, *New Atlantis* (1627; City, MT: Kessinger Publishing Co., 1999). Bacon's utopian vision is fascinating for a number of reasons. For one thing, as one of the original utopian works, *New Atlantis* casts an interesting light on many of the succeeding efforts, from Marxism to sustainability. It also illustrates very powerfully how the theological and technological drives became integrated early on in the development of Western culture, a unique coupling that had much to do with the eventual ascen-

dancy of Eurocentric culture. Along these lines, it is extraordinary how Bacon sees technological accomplishments (including manipulation of species) not just as a human accomplishment, but as a tribute to the glory of God, an interesting historical comment on those who view current genetic engineering as blasphemous ("humans playing God"). Those who are interested in tracing utopianism farther back might want to begin with the original eponymous *Utopia* by Sir Thomas More (first published in 1516), translated by P. Turner (London: Penguin, 1965).

4. World Commission on Environment and Development (Brundtland Commission), *Our Common Future* (Oxford: Oxford University Press, 1987). The original report that popularized the cultural construct of "sustainable development" is well worth reading on a number of levels. First, it is the authoritative source for the concept and thus for helping to identify and understand its core components, such as egalitarianism. It is also useful to read the report with several questions in mind: why, for example, is it considered necessary to invent this particular concept at this particular point in time? What are the social engineering goals implicit in the concept, and whose agenda do they represent? Such questions are, of course, appropriate to any cultural construct, especially one that is deliberately created for specific purposes, such as this one (Ian Hacking in his 1999 book *The Social Construction of What?* [Cambridge: Harvard University Press] provides a number of different examples). It is also worth noting again that cultural constructs are not necessarily bad things—indeed, they may be considered an indispensable way of ordering an otherwise unmanageably complex reality—but it is as well to be able to recognize them as such and to understand their dynamics.

5. J. E. Cohen, *How Many People Can the Earth Support?* (New York: W. W. Norton and Co., 1995). Cohen's book is both an excellent discussion of the complexities of quantifying the increase in human population over time and of attempting to establish a "sustainable" vision without asking basic questions about, for example, what kind of society, with what kind of income distribution, and what kind of base technology one wants to assume. His book makes it clear that the future condition of an anthropogenic world will not be a matter of gliding toward a utopian vision, but a product of choice, exigency, and contingency: creating cultural constructs does not avoid the difficult decisions and the ethical and rational responsibility that goes with them. There are many paths that can be taken, and the real question is whether humans will have the integrity to understand that or will avoid responsibility by lapsing into comforting, if blinding, ideologies.

CHAPTER 6

Faith and Science

It is perhaps ironic, but the more scientifically and technologically advanced a society is, the more the average individual in that society must rely on faith rather than on explicit rational knowledge—faith that planes will fly; faith that the Internet will connect her to friends, resources, and cyberspace generally; faith that a digitalized picture shows "the truth" rather than an enhanced or altered composite (and what is "true" if the changed image is more representative of what really happened?). Perhaps the oft-noted surge of belief in magic and religious stories in developed countries is not surprising, for if one is surrounded by apparently miraculous objects such as the Internet, and the technological and scientific basis for the object is as obscure as it is to most people, the differences between magic and technology potentially fade into meaninglessness.

This blurring of science and mythology becomes more than just academic when one turns to issues of environmental science and knowledge—if one asks, for example, how do we know what we know about "the environment," and what observations can we trust and not trust, and why? Our direct observations are necessarily at a local scale and, with complex systems that exhibit significant variability for reasons that are not intuitive, they may well be mistaken. If less fish are being caught off the South American coast, is it human overpredation, or is it El Niño and thus a result of natural oscillations in the atmospheric and oceanic systems? Global climate change remains in part contentious because there are no readily observable phenomenon that can without question be associated with the increase in carbon dioxide and methane in the atmosphere based on the observations an individual can make, and verify, personally. Loss of biodiversity is not an observable phenomenon in most places.

In general, virtually all environmental perturbations of any interest are extraordinarily complex and difficult to understand.

Environmental science and related disciplines such as toxicology are the primary means by which society seeks knowledge of environmental systems. In turn, many environmental discussions and political conflicts would be simplified if participants could agree on a basic set of factual assumptions. Frequently, however, this proves difficult. In some cases, this is because science, usually somewhat uncertain, is up against the certitude of ideology—in such cases, faith tends to trump reason. But in other cases, difficulty arises because the results of scientific inquiry are not accepted by all parties, usually because such results are perceived as normative and biased. This latter problem seems to be particularly acute with environmental science. Why?

To answer this, we must look at the scientific process itself. This is not uncontroversial. To some postmodernists, the entire scientific discourse is normative, a matter of beliefs and culture, not facts. On the other hand, to some logical positivists and scientists, the scientific discourse is the only road to Truth. Both of these extremes are oversimplistic, as a closer parsing of the scientific process indicates.

To take up the postmodernist argument for a moment, it is apparent that science as something that some people in a society do with their lives is a cultural phenomenon: it is thus in a broad sense normative, contingent, and reflective of the values and beliefs of its culture and historical period. Not all cultures would give science the authority it has in most developed countries today—Europe in the Middle Ages, for example, respected religious writings and statements more than observation and scientific analysis. It is also fairly clear that the choice of method, problem, and hypothesis are normative to some extent, reflecting broad social trends, the state and structure of the scientific discipline (Thomas Kuhn's famous paradigm), and the particular beliefs and training of the individual scientist. Thus, for example, an environmental scientist who is concerned about loss of species may pick a region to study and a particular hypothesis that reflect her or his belief. But at this stage also we begin to see the objective element of the scientific process, for regardless of the personal beliefs or ideology of the scientist, a hypothesis must be testable, and falsifiable, to be a legitimate scientific hypothesis. Thus, the statement "Hundreds of angels can dance on a head of a pin" is not a scientific hypothesis—but "Hundreds of species have gone extinct in this salt marsh" is.

Testing of the hypothesis through experiment and research, governed by strong practices and a robust scientific culture, is the most objective (or at least transparent) stage of the scientific process. Making up data, fudging results, reporting incomplete or skewed results—all are unacceptable. Similarly, there are disciplinary and cultural standards for the reporting of research; while disagreement over procedures is relatively common ("Was the proper statistical test used?"), misrepresentation of research results is considered scientific fraud. Many environmental scientists thus hold strong views about the state of the environment, but they also conduct their research and report their results objectively and relatively dispassionately. Interpretation of the results, either by the researcher or other groups, often follows and, depending on circumstances, can range from relatively objective to strongly normative.

But here the complexities of the natural systems with which environmental science must deal become relevant (see "Environmental Science and Postmodernism"). Much science is "reductionist"—that is, the scientist performs an experiment on a small part of a system that can be isolated and manipulated with control over all relevant factors. In this way, the actual testing of the hypothesis is not only apparent, but the experiment performed by the scientist can be replicated by any other scientist willing to take the time and follow the procedure. If the results cannot be validated by others, questions concerning the legitimacy of the initial findings immediately arise.

But it is not possible to study complex systems like the atmosphere or the Everglades just by doing reductionist science, for part of the problem is the need to understand the systems in all their complexity. And the distinction between the normative and the objective becomes very difficult to maintain when one moves from the relatively traditional, reductionist experiments that characterize most environmental science, to the nonreductionist, heavily computerized and model driven, nontransparent environmental science of global climate change simulations and biodiversity projections (see "Is Environmental Science Really a Science?" and "Is Environmental Science a Science?: Part 2"). These latter efforts, being highly complex and understood by only a few, generally make predictions that have significant potential impacts across social and economic systems, yet are inherently not testable by the usual reductionist scientific method. In short, there is a significant element of faith involved in accepting their results.

This is not a criticism of those that build or run such simulations, for they are studying complex systems at levels that necessarily are nonreductionist and require such techniques. There is no way to understand something like atmospheric dynamics at a global scale but to try to model it at that scale, and such models, if they are to exhibit the same behavior as the system one is trying to understand, are necessarily highly complex. Moreover, in general the practitioners make every effort to test the validity of their models (by, for example, running them using historical data and seeing how well they replicate what we know happened historically). The alternative to this difficult work—that policy in such areas as climate change ignore the knowledge that scientific research can develop—is clearly undesirable. How the best science can be done and contributed to policy dialogs under such circumstances remains a challenge to all concerned; in the global climate change dialogs, for example, the scientific community has been organized into the Intergovernmental Panel on Climate Change (IPCC), which brings together a large collaboration of relevant scientists in the preparation of periodic, authoritative reports.

But if a degree of faith is necessarily involved in accepting the results of such simulations, and if that faith is undermined by subjectivity in presenting the results and in extending them beyond the scientific scope of competency to make normative social and economic recommendations, there can be no surprise if those who disagree with the social recommendations also discount the validity of the science and technology of the underlying simulations. It should also be no surprise that this is especially true in environmental science, where it is frequently charged that the results achieved through nonreductionist scientific analyses are predetermined by activists, lawyers or policy mavens. For the environment is an emotional battleground and, at least in some cases, scientific results are often extended by scientist themselves, by NGOs and firms, and by partisan politicians. It is not uncommon, for example, to find an environmental scientist segueing from specific findings regarding changes in land use or historical climate records into general conclusions regarding significant loss of biodiversity or potentially disastrous climate change (or not, depending on the scientist's predilections), and then making policy recommendations. This potentially confuses two very different functions of the scientist: an expert in a specific area versus an ordinary citizen expressing an opinion on public policy. Sophisticated observers may note the difference, but frequently the media, and those groups that may have an interest in deliberately erasing the line between scientist and citizen, may not. Equally troublesome, when the scientist as ordinary citizen cloaks her or his opinions in presumptive scientific validity in contexts where significant resource allocation questions and prioritization across different value systems are at issue, the scientist preempts debate and silences dissenting opinion.

The result tends to be a blurring between environmental science as a relatively objective process that produces reliable information regarding certain phenomena, and normative positions dressed up with the patina of scientific objectivity. While some of the scientist's professional authority usually leaks into her or his policy recommendations—indeed, this is usually one reason activists encourage such comments—there is also an opposite flow: some of the normative flavor of the latter seeps back and undermines the veracity of the underlying science. The cost of losing this objectivity is high, for it means that there is less agreement on underlying data and phenomenon, and thus less opportunity to evolve appropriate policies. What should be a discussion about the best way to respond to identified changes in environmental systems instead becomes an ideological wrestling match about whether such changes even exist, making real progress even more problematic.

Identifying an environmental issue is relatively easy compared to the difficult question of what to do about it. What is clear is that simplistic prescriptions are probably inadequate, for the question of the authority of science, and particularly environmental science, arises not just from individual behavior that can be modified, but from much more powerful cultural trends that are not easily identified, much less changed. Moreover, it is the complex nature of the systems at issue, rather than choice, that requires the complicated tools of the computer modeler. Obviously, it remains important that individual scientists respect the line between their research and policy recommendations based on that research, for, as always, the integrity of science depends heavily on scientists themselves. But it is also important to increase the transparency of the environmental science and policy-making process and to make an effort to include those who may disagree with prevailing environmentalist views and visions. This includes not just individuals who may hold different opinions, but also studying disciplines and fields that historically may not have been part of the environmental dialog—trying to understand technological evolution and the role of technology in society, for example. For even the partial capture of environmental science by activists, while perhaps understandable from their perspective, in the long run undermines the ability of society to make the adaptations that in the end will be required.

Environmental Science and Postmodernism
(December 2002)

In the Czech writer Milan Kundera's novel, *Immortality* the worldly Bear tells the intellectual Paul that his very intellectualism acts in society to destroy the high culture it seeks to propagate: "I experienced it with my own eyes and ears after the war, when intellectuals and artists rushed like a herd of cattle into the Communist Party, which soon proceeded to liquidate them systematically and with great pleasure. You are doing the same. You are the brilliant ally of your own gravediggers."[1] A great line—but what does it have to do with environmental science?

To answer this, consider first the so-called culture wars between mainstream science and its postmodern critics. Oversimplifying, the latter holds that there are no totalizing discourses, no human belief structures that are fundamentally more valid than any others. Postmodernists thus argue that science has no more essential grasp on reality than English critics or shamans. Scientists, not surprisingly, reject this analysis, noting among other things the history of human development since the Enlightenment and the methodology of scientific advance, which requires validation of previous results. Somewhat more tartly, commentators favorable to science argue that postmodernists are merely frustrated intellectuals who, having steadily lost power over the past few centuries in an increasingly technocratic society, seek to regain it by diminishing science, the most dominant intellectual theme of modern culture. To regain their wonted intellectual authority, postmodernists must bury science.

This is a complex argument, and it is unnecessary to become bogged down in details. It is sufficient for our purposes to make several observations. First, science, including its implementation in technology, remains the dominant discourse. This dominance is supported not only by the operation of market capitalism, which rewards technological innovation, but also by the cultural power of science. Consider the global climate change negotiation process, which continues to pretend it is only about science when, in fact, it is profoundly about values (any response to global climate change inevitably favors certain paths of human and biological evolution, and the values implied in such paths, over others). It is the mark of a dominant discourse that its language is used even where inappropriate to the underlying subject matter.

---

[1] M. Kundera, *Immortality* New York: Harper Perennial, 1999.

The second, and somewhat contrary, point arises from the branch of environmental science that deals with highly complex systems that cannot be understood by replicable experiment, such as climate change and the like. Call this "nonreductionist environmental science," and note that in many cases the tools used to study such systems, such as large computer models, are neither transparent nor verifiably objective in the usual scientific sense. Moreover, perhaps because of the urgency felt by many who are involved with nonreductionist environmental science, there is a strong tendency to conflate objective observation based on science with personal normative positions. These statements of concern are in turn picked up by powerful interest groups, especially environmental NGOs, and are used to drive continuing high levels of public concern about the environment (and continued funding for environmental groups). Historical examples might include the predictions by respected ecologists in the 1970s that human population growth would create worldwide starvation and disaster by the 1990s, or the long series of predictions of major resource shortages by environmental scientists. More recent examples might include the new field of sustainability science, which by its title validates the existence of the very problem that its research presumably is intended to find. Expressions of strong concern have in some cases played the salutary role of identifying important issues requiring attention (as in the case of acid rain or lead in gasoline), and in some cases have been simply wrong: what is important is to separate to the extent possible the scientific underpinnings from the public campaigning.

Does it matter if the normative and the objective are indistinguishable in nonreductionist environmental science? Note first that an inability to separate the normative and the objective does not mean that the underlying phenomena are unreal. It just means we cannot rely on the data and statements as reasonably objective. But by giving over scientific objectivity and methodology, nonreductionist environmental science, and its partisan use by environmentalists, risks eventually undermining the validity and authority of the scientific discourse in environmental issues altogether. It thus changes the basis of environmental debates from relatively objective discussions of the state of the real world, to normative conflict between environmentalists and their critics—and risks positioning environmental science as just another transitory belief structure devoid of objective content. In this area at least, we seem perilously close to validating the postmodernist critique of science as nothing more than another normative power play by specific social interests—and thus science allies with its own gravedigger.

Is Environmental Science Really a Science?
(August 2002)

That challenging and seemingly almost insulting question (at least, to an environmental scientist such as myself) arose in a seminar last year. I had, in fact, managed to successfully ignore it until another recent meeting raised it in a pointed way about the related field of industrial ecology. Is the "purpose" of industrial ecology to support sustainability, or is it to conduct relatively objective research that may be used, among other things, to improve the environmental impact of various industrial and economic patterns? Is the "purpose" of environmental science to perceive and validate new knowledge, or is it to impose normative institutions and futures on society (for example, by forcing society to be "sustainable")? Is "sustainability science," announced in *Science* magazine and now centered among other places at Harvard University, a science or an ideology—and does it matter? Is there a "science" in "environmental science," and, if not, does it matter? A couple of short columns, while not providing definitive answers, can perhaps reach some worthwhile conclusions.

To begin with, we may reject the traditional view of science as a purely objective discourse. This view was pretty firmly laid to rest by Thomas Kuhn in his classic *The Structure of Scientific Revolutions*, where he pointed out that science, like any human activity, reflects its time and place in many ways. Thus, for example, environmental science has flourished in the last few decades because "the environment" has during that time become important for society in ways it never was before. But there is an important methodological difference between science and other discourses that does make it unique: scientific hypotheses, theories, and facts must be falsifiable, and scientific disciplines put a high premium on being as objective as humanly possible. In the words of Bertrand Russell, "In the welter of conflicting fanaticisms, one of the few unifying forces is scientific truthfulness, by which I mean the habit of basing our beliefs upon observations and inferences as impersonal, and as much divested of local and temperamental bias, as is possible for human beings."[2] Thus, science does not preclude normative activities by scientists—but it does at least put a burden on them to separate their normative from their scientific activities, especially as regards methodology and data generation.

This insistence on differentiating the normative from the objective is

---

[2] B. Russell, *A History of Western Philosophy*, New York: Simon and Schuster, 1972, 836.

especially important because, in modern society, the amount of information is so overwhelming that individuals, no matter how educated, are unable to test for themselves all scientific statements. Thus, they must take a considerable amount of the world on faith. I haven't the faintest idea how to design a 747, or how an Internet router works, or whether a climate scientist's model results are predictive. But I take them on faith and do what operational tests I can manage: the 747 flies, the Internet delivers information to my computer, and the environmental science community supports the conclusions of the climate scientist. But it is faith and minimal observation, not validated knowledge, that supports my belief in them. Ironically, in the quintessential age of reason we have come to rely on faith to an unprecedented degree.

Faith may, however, be justifiable. Much environmental science is reductionist: based on specific tests of particular materials, for example. Such science can be replicated, and in part I am entitled to believe it because, if it is fraudulent, it will be falsified when people cannot replicate the claims. Thus, for example, a claim by one researcher that salmonoid species prosper in water heavily contaminated with heavy metal can be rapidly falsified by others performing the same or similar research. In other words, the scientific process and its institutions—including peer review and publication in such journals as *Environmental Science and Technology*—operates with reductionist science to validate its objectivity. I am comfortable in asserting that this type of environmental science is "scientific": where results are based on bad technique, fraud, or ideology, the methodology of science will eventually expose it. The field may be culturally contingent (environmental science is popular now because the environment is), and research projects may be chosen for normative reasons (I study toxic metals because I think there is a problem), but the results and the data are, thanks primarily to the scientific method, valid. Can I, however, have the same confidence in the second fundamental type of environmental science, which I will call nonreductionist environmental science (NES)? That is the next question, and the next column.

Is Environmental Science a Science?: Part 2
(September 2002)

In last month's column, we began contemplating this difficult question by noting that much environmental science is reductionist, and thus supported by traditional scientific methodology. But there is a second type of environmental science that involves highly complex systems, where the basic knowledge base is not yet mature and stable, or indeed even adequate, and where the only tools available are themselves necessarily highly complex and far from transparent (for example, computer models). Let's call these areas of study nonreductionist environmental science, or NES for short. NES can involve everything from difficult epidemiological studies trying to identify subtle effects in large populations with numerous potentially confounding factors, to estimates of loss of biodiversity, to large global circulation models that predict the effects of climate change. Importantly, in almost all cases, this type of environmental science cannot be falsified or replicated, is not testable by experiment, and is not transparent or intuitive to the uninitiated.

Because it cannot be falsified, NES begins with an inherently more difficult burden of consistently demonstrating objectivity, rather than deliberate or inadvertent biasing of results for ideological reasons. This burden is all the heavier because the policy recommendations that are drawn by environmentalists and environmental scientists from NES—for land use, for reduced economic activity, for limiting consumption and development—are often profound, and frequently at odds with traditional policy assumptions such as the desirability of economic growth. Under such circumstances, those adversely affected by such policy recommendations are likely to demand a high level of proof before they acquiesce to them. Their skepticism is liable to be magnified by their recognition that many environmental scientists are drawn to their field through a powerful self-selection process that frequently includes a strong sense of pending environmental disaster, and thus these scientists may already have a worldview that makes them prone to conflate the normative with the objective.

It does not help that the results of NES almost always feed into a highly charged political discourse in which environmentalists can use results as primary ammunition to advance their positions and policy prescriptions as privileged over any others—jobs, economic performance, development, or the like. This process tends to reposition science as a subfield of policy. For example, the deep green theologian Thomas Berry writes, "Ecologists recognize that reducing the planet to a resource base for consumer use in an industrial society is already a spiritual and psychic degradation. . . . To the ecologists, the entire question of possession or use of the earth, either by individuals or by establishments, needs to be profoundly reconsidered."[3] While practitioners of NES obviously do not control what such commentators say, the effect is nonetheless corrosive of perceived scientific objectivity.

Perhaps more problematic is the perception that NES performed by environmental regulators and governments, as much of it is, is heavily biased toward supporting preconceived policy goals. Thus Mark Powell, in a Resources for the Future study, quotes an EPA official to the effect that "it's hard to avoid being perceived as an intellectual gadfly or snob when you understand your mission to be cajoling [EPA] program offices into taking science seriously and not playing games with the numbers to prop up a political position."[4] And just recently *Nature* reported that an independent scientific review by Germany's science council of the Wuppertal Institute— which is one of the leading environmental research and policy institutions in Europe, and certainly one of the most well-known—"slammed" it for "selective" results and recommended that it no longer receive public money unless the quality (objectivity) of its scientific research improved.[5]

It is probably too strong to suggest that there is a crisis of faith regarding NES, although the travails of the Kyoto process suggests that possibility. But it is not too strong to recognize that as NES continues to expand into industrial ecology, sustainability science, and the like, it increasingly jeopardizes its validity to the extent it fails to rigorously observe the critical difference between scientific and unabashedly normative approaches, especially in highly contentious public debate. Existing policies and practices, like all social institutions, have a great deal of inertia and powerful forces supporting them; changing such policies and practices will require equally powerful challenges. The more NES is understood to be normative and partisan rather than objective—the more it is seen as ideological rather than scientific, especially in method—the less likely such change will be.

---

[3] T. Berry, "The Viable Human," in M. E. Zimmerman, J. B. Callicott, G. Sessions, K. J. Warren, and J. Clark, *Environmental Philosophy: From Animal Rights to Radical Ecology*, 3rd ed., Upper Sadde River, New Jersey: Prentice-Hall, 2001, 175–184.
[4] M. Powell, *Science at EPA: Information in the Regulatory Process*, Washington, D.C.: Resources for the Future, 1999, 136.
[5] *Nature* 417:678.

Annotated Bibliography

1. C. P. Snow, *The Two Cultures* (1959; Cambridge: Cambridge University Press. 2001) and T. S. Kuhn, *The Structure of Scientific Revolutions*, 2nd edition (1962; Chicago: University of Chicago Press, 1970). These are two classics of the sociology of knowledge. They should be read simply because they are so fundamental to all that follows them. Snow's work has become synonymous with the deep cultural differences (and ontological differences, for that matter) between the scientific and technical versus the social science and literary critic ways of knowing. Somewhat ironic, actually, as his real point, much in line with sustainability, was to enlist the scientific and technological community in addressing poverty and lack of economic development. Kuhn's work was gently revolutionary: revolutionary in that it overthrew the reigning image of science as a methodical advance of progress founded only on rational analysis of data and testing of objective hypotheses, but gentle in that it recognized the normative character of some parts of the scientific process (such as hypothesis generation) without denying the relatively objective components of that process (such as conducting the research itself). The balance between revolution and incremental critique depends somewhat on the interpretation of the observer . . . so you should read it and decide for yourself (reflexively, it is amusing that, in good postmodern style, the interpretation of the reality of Kuhn's work is in the hands of the critic).

2. J. Lyotard, *The Postmodern Condition: A Report on Knowledge*, translated by G. Bennington and B. Massumi (1979; Minneapolis: University of Minnesota Press, 1984). Lyotard's book is a very readable postmodernist discussion of knowledge, covering most of the major themes of the genre: the ability of a "dominant discourse" such as science and technology to establish power relationships that define social reality; the fundamental relativism among discourses, meaning that none are inherently "better" than any other; the subjectivity of knowledge. While the Marxist strain is evident in this analysis, it is always useful to get an outside critique of a powerful cultural system; you may not agree, but you at least need to think about why you disagree, which is a good exercise.

3. M. Bookchin and D. Foreman, *Defending the Earth: A Dialog Between Murray Bookchin and David Foreman* (Boston: South End Press, 1991). Foreman is best known as the founder of Earth First!, while Bookchin is a leading social ecologist. Their dialog is interesting not just because it is a good introduction to nonmainstream environmentalism, but—perhaps because it is not mainstream—it illustrates some of the foundational themes of the environmental discourse more clearly than less ideological sources (of

course, while the message is interesting, the reader should remember that many environmentalists regard the direct action of groups such as Earth First! and the Earth Liberation Front as inappropriate and counterproductive). Additionally, the astute reader can catch strands in this dialog of the "two cultures" phenomenon ("in the wild" as it were): is the environmental absolutism that Foreman espouses a reflection of objective observation, or is it—indeed, is all environmentalism—a cultural phenomenon as someone like Bookchin might argue? Finally, the very title itself is revealing, and you might want to consider whether it is the Earth as an evolving complex system with no teleology that is being (or needs to be) defended, or a particular vision of the Earth held by a select few.

4.   B. Lomborg, *The Skeptical Environmentalist* (Cambridge: Cambridge University Press, 2001). An interesting contrarian review of the environmental discourse and some of its most prominent claims, raised to a cultural event in itself by the outraged reaction of some in the environmental science community. With staggering hubris, for example, *Scientific American* titled its response "Science Defends Itself against *The Skeptical Environmentalist*," as if (a) any publication could speak for all of science; (b) whatever Lomborg had said was so beyond the pale that all of science was arrayed against it; and perhaps most interesting for a scientific publication, (c) no dialog or discussion, merely complete and total rejection, was appropriate (some scientists argued that the book should never have been published in the first place, thus encouraging an unfortunate perception that, at least in science, censorship is preferable to debate). The fight over Lomborg's book goes on—even including a finding of scientific misconduct by the Danish Committees on Scientific Dishonesty, which in turn found itself accused of gross misconduct since its finding apparently relied not on any independent evaluation, but principally on the *Scientific American* review. On the one hand, this is great theater; on the other hand, serious effects include a strong signal that dissent in environmental science will be punished and obvious personal impacts on the individuals involved. What is the honest reader to do? First, read Lomborg; second, read the criticism; then decide for yourself. And think of the book on two levels: first, and perhaps ironically less important, for what it says and what its critics respond on a factual basis; and second for what the environmental science community's response, and subsequent events, say about the separation of ideology and objective science in this particular case.

CHAPTER 7

Complexity: The New Frontier

If there is one concept critical to understanding the anthropogenic world, it is "complexity." But there are no easy answers to the obvious follow-up question: what is complexity? Indeed, like many modern concepts, complexity has rapidly entered popular discourse and has thus attained a state of ambiguity somewhat similar to that previously discussed regarding sustainability. But, as in that case, one can at least begin to identify important aspects of the concept (and read some of the books mentioned below), which may not support a formal definition but will at least provide a much better intuition of what complexity is and how it behaves.

To begin with, it is apparent that complexity arises only with systems, collections of agents that interact with each other in some way. Such systems can be either *simple* or *complex*. These two classes of systems behave in quite different ways. Moreover, they are perceived by people in quite different ways as well, for our psychology predisposes us to think in terms of simple systems. That is, people tend to conceptualize in terms of relatively few variables whose interrelationships are easily understood and which are displayed over a short time period. That is one reason why people react strongly when a stream is polluted because of releases from an upstream factory (clear cause and effect), but less so with climate change or changes in biodiversity. We do not intuitively understand phenomenon such as climate change or the operation of the global economy or complex technologies: there are too many variables interacting in too many ways over too many different temporal and spatial scales for us to grasp the system as a whole.

Unfortunately for our intuition, virtually any nontrivial element of the anthropogenic world is highly complex. More than that, the world is getting more, not less, complex (see "Tribalism versus Globalism?"). In turn, this complexity comes in many different flavors. It may be static complexity, where just the map of agents and their interactions are baffling. An example might be a map of the hubs and links composing the Internet, a map that remains perpetually incomplete precisely because of the complexity of the underlying phenomenon. It might be dynamic complexity. Peter Senge, in teaching business students about complexity, uses a very simple model called "the beer game," consisting of a retailer, a wholesaler, and a marketer of beer. Despite the obvious static simplicity of this model—it only includes three decision nodes—once a few dynamic elements are introduced into it (time lags between the responses of the wholesaler to increased sales at the retail level, for example), the game becomes almost impossible for participants to rationally optimize. Finally, most systems characterizing the anthropogenic earth also include what we can consider as human complexity: the contingency that history and experience show constantly arising from the highly reflexive and coupled human systems within which we operate. This last flavor of complexity raises messy questions, such as the existence, nature, and limits of free will, which, having been argued over for at least two thousand years, we can justly assume are quite complex in themselves.

Indeed, what a "system" is for a particular purpose is not necessarily as obvious as common usage might suggest. If we define systems as groups of interacting, interdependent parts (agents, or subsystems, or lower hierarchies within the system under consideration) linked by exchanges of energy, matter and/or information, we are also implicitly defining a boundary, so that we can say that some things are inside the system, while others are part of its outside environment (or "endogenous" versus "exogenous" in more formal language). It is important to recognize the somewhat arbitrary process involved in drawing such boundaries, for it is characteristic of complex systems that we define the relevant boundaries in part by the query we pose. For example, if I ask a question about New York City crime statistics, I am implicitly creating a boundary that reflects jurisdictional structures, in this case the five boroughs. On the other hand, if I ask a question about the adequacy of New York City's water supply, the boundary of the system includes most of New York State, which has been tapped over centuries to provide water for the city. And if I ask a question about the city's commerce in clothes, I include most of the world: the network of transportation, financial, and other systems required to design, make, and ship the clothing to the city.

In general, a good boundary is one that internalizes as much of the relevant interactions as possible, while at the same time minimizing the relevant interactions of the system with its environment, thus facilitating analysis. Boundaries, however, always reflect a subjective element. Put another way, the boundaries of a complex system are as much a product of the observer (or inquirer) as of the "real world out there," a limitation that should be borne in mind when the results of any such analysis are applied beyond the implicit boundaries of the original inquiry. To take our New York City example, an analysis of the water system based on the jurisdictional boundaries appropriate to understanding the crime patterns of the city would be flawed from the beginning, for each inquiry necessarily implies a different set of relevant agents and linkages.

What are some of the most important differences between simple and complex systems? Perhaps most obvious is linearity. With simple systems, if an input changes by a certain amount, the output changes by an equivalent and predictable amount, and the change in output is relatively comparable over the entire range of inputs and outputs. Complex systems, on the other hand, are generally characterized by nonlinear interactions between the system components. If an input changes by a small amount, the output may change dramatically, depending on the internal system dynamics. For example, a salt marsh may look relatively unaffected for a long time as the pollution input increases, but at some point just a small increase in pollution results in a significant degradation in the salt marsh ecology. The relationship of pollution to salt marsh response is nonlinear.

Similarly, simple systems tend to be intuitive, in the sense that their operations are all apparent. In practice, this means that there are no significant spatial and temporal lags, discontinuities, or limits and thresholds. With complex systems, on the other hand, spatial and temporal lag times cover a broad range, and system behavior cannot be projected simply by focusing on predictable near-term trends. Such systems are also known for discontinuities, where a small change in input creates a discontinuous change in output, and they display a number of limit and threshold effects. The result of these dynamics is that complex systems, unlike simple systems, generate internal behavior seemingly "out of nowhere" as a result of their internal dynamics.

One example of such emergent behavior is the crash in the U.S. stock market on October 19, 1987, known as Black Monday, when the Dow Jones Industrial Average dropped 508 points, losing more than 22 percent of its value. Although this remains the record for a single-day loss on Wall Street, it was followed by a record one-day gain the next day of more than 100 points and a subsequent jump of more than 185 points on October 22. Naturally, such huge swings in value over a few days time generated a lot of postmortem analysis, which identified a few major causes: the lack of real-time synchronization between stock markets and derivative markets (indexed options and futures markets); the relatively new computer trading programs, which substituted internal thresholds and values for human judgment; the market's inability to remain liquid at unexpected high volume (in part generated by the computer trading systems that created unanticipated positive feedback loops); unexpected interactions of routine financial data (especially regarding trade deficits) with the above market dynamics.

What does all this mean? It means that the single largest drop in market value in the most sophisticated stock market in the world resulted not from any external financial cause, but from the internal dynamics of a complex system that contained powerful feedback loops, threshold effects, and discontinuities that were not recognized and certainly not understood. The drop, in other words, was a result of internal dynamics, not any particular outside act or event that changed the market. It is a now classic case in the behavior of complex systems.

Moreover, with simple systems it is usually easy to determine cause and effect relationships. Complex systems, however, tend to be characterized by complex feedback loops and high degrees of interconnectedness that make it difficult to distinguish cause from effect. Indeed, a common analytical mistake in evaluating complex systems is to impose a cause-and-effect model on the system's behavior, a subtle but seductive error because of the human tendency to think in such terms. This can be a particularly tricky issue given that a complex system is defined in large part by the nature of the query one poses to it: it is entirely possible to look at the same set of interrelationships and conceptualize it as a simple or as a complex system, depending on the purpose for which the system is being considered. "New York City" will mean very different things depending on our purposes for using the term, and some of these meanings will imply simple systems ("What is the weather in New York City today?") and some complex ("What is the role of New York City in global financial flows and their impacts on global ecosystems?"). In either case, the underlying structures, functions, and relationships are the same, but the parts of the system that are relevant to the particular query differ.

Simple systems tend to have a defined, static, equilibrium point toward which they move regardless of starting point and independent of time. A marble in a bowl, for example, will move toward the bottom of the bowl regardless of where on the bowl's surface you start it or when you let go of it. Complex systems, on the other hand, usually have no such easily defined equilibrium point. Rather, they tend to operate far from any equilibrium point(s) in a state of constant adaptation to changing conditions—cities, multinational firms, and salt marsh ecosystems being examples. Indeed, the old view that ecosystems should be understood in terms of ecological succession (communities replacing or succeeding prior ones in the same location) progressing toward a specific stable endpoint, which is then maintained forever through time, is being replaced with a model where the community is seen as constantly shifting in response to its internal dynamics and a (constantly evolving) external environment.

With simple systems, one can predict and understand larger systems just by adding up the properties of the constituent subsystems: the composition of a fruit salad can be understood simply by adding up the properties of the different fruit cut up to make it. Complex systems, however, are generally characterized by emergent characteristics that appear only as one moves from subsystems to higher levels in the system hierarchy, and such behavior cannot be obtained simply by summing up the parts' contributions to the whole. Thus, for example, a nerve cell exhibits behavior that could not be predicted simply by knowing its chemical constituents; human consciousness would be impossible to predict simply through the study of nerve cells; and the structure of society cannot be predicted just by knowing a single individual. The behavior that emerges at each higher level is qualitatively different than what could be predicted by considering the characteristics of the lower levels alone.

Similarly, simple systems do not evolve. If disturbed, like a marble in the bowl, they return to equilibrium. In contrast, complex systems evolve in response to changing boundary conditions and internal dynamics. Moreover, they are "path-dependent," in that their past states limit the future paths along which the system can evolve. Once you mix the cake batter and bake it in layers, you can no longer just remove the flour: mixing and baking the cake precludes that path. Thus, for example, the United States cannot go back to the way it was in the 1920s; the cultural, economic, political, and technological evolution that has occurred since then mark a path that cannot be reversed. Put another way, there is no equilibrium point toward which U.S. history would return if it were not continually perturbed.

This has significant implications: for example, it is this characteristic that explains why the Everglades can never be "pristine" again. Its future states are functions of its past states, and those past states are heavily influenced by humans. The paths that are available for future design and management of the Everglades have been determined by the decisions, activities, and dynamics that have already occurred in the history of the region. That does not mean that certain values—for example, preservation of avian diversity—cannot be supported as part of the design of the system, but it does mean that restoring the Everglades to its prehuman condition is simply not doable.

Not only is the evolution of a complex system path-dependent, but there are no optimal results, for there is also no assurance of optimal efficiency in an evolving complex system (nor, indeed, is it necessarily clear what "efficiency" may mean in such a context). Indeed, in many ways the idea of "optimality" is too simple for complex systems, because their interconnectedness means that any effort to optimize one element of the system will almost inevitably cause other connected parts of the system to shift in unpredictable ways.

Taken together, these characteristics of complexity have several implications for an anthropogenic world. For one, they indicate that trying to predict sustainability or any such global characteristic of a complex and evolving system is liable to be virtually impossible (see "Complex Systems and Rebound Effects"). Moreover, they strongly suggest that the tools that many of us use to perceive, and function in, the highly integrated human/natural world we live in—especially the ideologies and cultural constructs we are so fond of in environmental dialogs—are oversimplistic, limited in value, and lead to misleading perspectives (see "Marx, Environment, and Complexity").

For example, consider the common assumption that the global climate change negotiations process is the main means by which humanity can respond to global climate change (and thus that the failure to ratify the Kyoto Treaty would have demonstrated a lack of response). This amounts to an assumption that global climate change is technically a simple system that can be managed by a centralized, explicit, control mechanism like a treaty. If that assumption is right, it means that we understand what the impacts of such treaties—on natural systems, on economic, political, and cultural systems—will be. Most people would not embrace this conclusion, and, indeed, one would not expect any form of centralized management or control system to be effective in a truly complex system. If nothing else, the information content and dynamics of such systems preclude it. And when one turns from the simple system model, one finds in the case of global climate change significant evolution in many decentralized forms, operating at very different temporal and spatial scales in many different ways, most of which are not explicitly coordinated but rather self-organize in unpredictable ways.

For example, most firms in affected sectors—insurance and financial; energy production and distribution; biotechnology, agriculture, and biomass; automotive, and so on—are actively researching and exploring alternatives. The motives may be purely defensive, but no firm with any capability is going to be blind to where the science and the politics increasingly point. NGOs continue to explore alternative strategies. Engineers begin talking about adaptive technologies, and technological evolution begins to reflect concerns about global climate change (examples abound: techniques for sequestering carbon from coal-fired power plant emissions; telework and virtual office networks; hybrid automobiles; renewed interest in nuclear power; hydrogen economy initiatives). Academics increasingly study global climate change from its various perspectives. Religions begin to think about how environmental perturbations such as global climate change fit within their traditions.

The change in the world around us over the past decade or so regarding global climate change, taken as a whole, is extraordinary—and for most people, even (especially?) those most heavily involved, is almost invisible. Such adaptation is indeed complicated, but that is the way evolution occurs in complex systems. It is not only that our mental models, tied as they are to the past, prevent us from seeing the adaptation occurring all around us. The difficulty is exacerbated in practice because, in a world that is highly complex socially, technologically, and economically, many such adaptations and technologies may well be invisible to all but those directly involved.

All of this is not to say that we should not use simplifying structures such as mental models—we must do so to make sense of the world, for language itself is a critical simplifying structure—but it does argue for doing so consciously and with great care (see "Changing of the Guard"). It is critical in a complex world to recognize the limitations of our intellectual tools and our disciplines, a point that sounds trivial and would be except for the obvious fact that it is ignored so widely in practice. In particular, it is important to beware the false certainty of ideology, which is liable to the dual failure of being not only oversimplistic, but also an application of past verities to a future that in most cases promises to be radically different.

More positively, it is obviously worthwhile to place more emphasis on understanding critical behaviors of complex systems, rather than trying to completely define them or determine their sustainability. One theme that rises to the top of such a research agenda is resilience, or the ability of a system to absorb insults and challenges without significant change or degradation (see "Resilience in Complex Systems"). Even where the dynamics of a particular system are not completely understood, it is still possible to design in, or create conditions for, resilient responses by the system to unanticipated challenges. Thus, for example, supporting an ecosystem that has several species for each functional area (such as shredders in a freshwater aquatic ecosystem) can, regardless of other benefits, help create resilience: even if an unanticipated event—say, a disease or otherwise minor change in environmental parameters—eliminates one species, the others can continue to fill the ecological niche and provide the requisite functionality. Ignorance does not mean powerlessness, nor that there are no preferable options, no ways forward. We may not know enough to do the best, but in most cases we certainly know enough to do better—and design for resilience, as an important element of earth systems engineering and management practice (further discussed in chapter 9), is an example.

Tribalism versus Globalism?
(November 2001)

One of today's great myths, held by radical environmentalists among others, is that tribalism (a.k.a. community) and globalism are locked in conflict. Like many mythic confrontations, there is a grain of truth in this worldview; but it is also a serious distortion, one that increasingly leads to violence and intolerance. It is thus worth exploring, even if only in a short column.

The implicit assumption held by many activists, environmental and otherwise, is that the world used to be a collaboration of more or less separate communities and countries, each with its own culture and economy, a state that was both culturally diverse and environmentally more benign. This Edenic vision is usually complemented by the belief that indigenous societies are culturally and technologically sustainable, as opposed to industrial culture, which is not, and that in the inevitable conflict between them, the latter is deliberately destroying the former.

This worldview is seriously flawed for a number of reasons, most of which are discussed in greater detail elsewhere (the September 29, 2001, *Economist*, for example, dedicates its center-section survey to "Globalization and its Critics"). But I wish to focus on one fundamental flaw that characterizes this mental model: an inability to appreciate dialectic (for those with a Marxist bent) or, put differently, an inability to appreciate human complexity and the continuity of change it necessarily entails.

The basic, usually implicit, assumption of the tribalism versus globalism argument is that the world is static; therefore, conflicts occur in the context of a zero-sum game. Someone must win, and someone must lose. Thus, if globalism wins, tribalism must lose, and vice versa: there can be no better outcome. This assumption is, however, simply wrong as regards an evolving complex system, for which conflict is not necessarily bad, but is the very engine of growth. Thus in this case, the world is growing not just more tribal, or not just more global, but both at the same time. It used to be that the idea of culinary diversity in most American cities was hot dogs versus hamburgers; now, it is a small town indeed that does not have numerous choices between Vietnamese, Thai, Afghan, Indian, Mexican, and perhaps many other cuisines. Are there fast-food restaurants in Tokyo? Sure—and, conversely, Japanese restaurants are here in the United States. Geographic communities may be less strong, but they are rapidly replaced, indeed augmented, by Internet-based communities of interest. People in a village used to know perhaps tens of neighbors; now, in developed countries both in person and online, they interact annually with thousands. More fundamentally, the entire structure of postmodernism, with its pastiche of different times, places, and cultural patterns, is a reflection of a growth in complexity that springs from a world where most people cannot live without knowing at least something about other societies, other places, other beliefs. International governance systems used to involve only nation-states; now firms, NGOs, and interest groups participate in a much more multidimensional and confusing, still nascent, process.

Whether these changes induce a defensive and intolerant reaction or tolerance and an appreciation of difference, the world is increasingly complex for most people. And it evolves. It is not a zero-sum game, tribalism or globalism; it is both. And, in different and new ways, both are growing more robust. The explosive growth in numbers of NGOs in the last twenty years is not a story of globalization; it is a story of new tribes created by, and supported by, globalization of underlying technologies and languages. It is a story of community and globalization coevolving—a story of increasing complexity. It is not that there aren't cultures and tribes that die out, for any creative, evolutionary process necessarily progresses by subsuming or growing beyond some elements, while generating others. The intellectually honest response is not denial, but construction of a more humane world where such losses are cushioned and unique cultural and linguistic information preserved against loss.

This process of increasing complexity brings with it greater choice and liberty—slowly and sporadically, no doubt, but it is there, from the trivial examples of exotic cuisine becoming widely available to the more fundamental ability of individuals to join with NGOs representing their beliefs, and to thus greatly augment individual power. It is no doubt convenient to adopt the fiction of a fundamental conflict between tribalism and globalism. But it is not just wrong; it is an insult to human creativity and freedom.

Complex Systems and Rebound Effects
(September 2001)

Telework, done right, is almost a poster child for the "triple bottom line" (TBL): it offers a number of economic, social, and environmental benefits, which are documented, and challenged, in detail elsewhere. In this column, I want to focus on some interesting environmental nuances—anecdotal, of course, but counterintuitive and unexpected—that, but for personal experience, I would not have noticed.

To begin with the background: As befits one of the world's leading telecommunications firms, AT&T has a world-class teleworking initiative. More than a quarter of all AT&T managers telework at least once a week; further, more than 10 percent have no assigned workspace in any AT&T building, and thus work in virtual offices—known as "going VO" among the cognoscenti. As one of those responsible for this program, I have gone VO as well—there is nothing like experience to help one understand the issues that arise from new programs such as telework on an institutional scale.

To begin with, I have found that working out of a home office not only does not encourage me to take more side trips for shopping and such than I did when I commuted daily—it discourages the habit of driving. The patterns tend to shift from a mental model where one must drive to everything (beginning with work) to one where the world comes to you, in the form of e-mail, and Internet and intranet services. Of course, I still shop—but I am doing less errand driving than I did when I commuted. For AT&T as a whole, our calculations, based only on avoided commutes, indicate that teleworking avoids 110 million miles of unnecessary driving per year, saving approximately 5.1 million gallons of gasoline and the emission of some fifty thousand tons of carbon dioxide. My experience, however, suggests that—especially as the new mental model becomes routine—those numbers may in fact be understating the benefits of telework. The implicit mapping of the world has shifted in some interesting ways.

Perhaps that effect could have been predicted (though I know of no one who did so). What I would not have known before actually going VO is the significant resultant decrease in laundry. Right. Laundry. And hold the jokes. This summer, for example, I have generally worn a swimsuit to work (why not? The only ones that see me in my home office are my computer and my cats and—so far—they don't seem to care one way or the other). The combination, therefore, of less public time encouraged by the new mental model, with the significant reduction in dirty clothes enabled by a nonoffice environment, is that I have gone from generating about two laundry loads a week to perhaps one every two weeks. That's a lot of hot water and soap saved in the course of a year. Moreover, I am now home so I can line-dry the stuff, which saves even more energy. And this all results not from a negative and authoritarian "give up all consumption and live like a peasant" approach, but from a lifestyle that I find much preferable to the commute/office routine. In personal affairs, as in international environmental negotiations, one catches a lot more flies with honey than with vinegar.

So, from this limited experience, several observations. Never give in to the hubris, or ideological certainty, that you understand a complex system. You don't. There is no substitute for experimentation and experience. This requires, however, a conscious effort to understand many activities as experimental and to continually observe and learn from them, a posture that is unfortunately not widely practiced. A much more sophisticated emphasis on learning as an ongoing social function by all parties—firms, governments, NGOs—is thus highly desirable. Barriers to such an experiential posture are formidable: among other things, this mindset is to some extent incompatible with the unthinking acceptance of popular ideological approaches such as the precautionary principle that, if taken strictly, become simply a prescription for stasis and thus continuing social inequality and environmental degradation. Moreover, a bias toward experience and observation, as opposed to ideology and myth, is particularly difficult in the environmental policy arena that is frequently characterized, at least implicitly, by ideological posturing and a concomitant disinclination to observation and objective assessment of data. In practice, many groups involved in such issues already "know the answers" and observation of any kind is treated as dangerous, because it might disprove powerfully held beliefs. In an increasingly anthropogenic and complex world, that is an increasingly indefensible posture.

Marx, Environment, and Complexity
(October 2000)

Some people may wonder why this column, written by an industry person, refers so often to the thoughts of Marx, which at least in recent history have formed the basis for the most profound challenge to market capitalism. The answer is easy and explains this month's column: Marx was a brilliant thinker, and he saw clearly the human costs—as well as the benefits—of the capitalist system evolving around him. We can learn a lot from where he was right or, as in this case, wrong.

It is often said that every philosophic system has a nightmare potential that turns it from good to bad—in Marx's case, his political theory, intended to enable human freedom, turned into a mechanism for despotism and tyranny. This resulted from his serious misunderstanding of the nature of the state. If the state is just an epiphenomenon of class, it goes away when the workers win. That is what Marx hoped. But suppose the state is more than that—an independent power on its own, a necessary creature given the complexity of Enlightenment culture and economies. Then not only is Marx wrong about it going away, but his vision of a workers' paradise, where each gives according to her ability and receives according to her need, becomes a fantasy. Instead of paradise, one has the looming evil of Stalinism as the state usurps and perverts Marxist theory.

No prize for guessing which vision was right. But why? The answer can be a powerful teacher for modern environmentalism. There are many critiques of Marxism, but they generally miss one fundamental mistake: he misunderstood complexity. In essence, he thought an increasingly complex economic, cultural, and political system—bourgeoisie capitalism—would spontaneously revert to a simpler state (agrarian anarchism, perhaps?). Of course, it did not: absent some kind of collapse (in this case, of population, culture, or economics), such systems instead continue to evolve greater complexity. I call this the "Marxist complexity fallacy."

This becomes an important guide for environmental policy, especially for selecting endpoints. A vision frequently expressed, especially by deep greens, is a world where humans are gathered into small, urban islands, and the rest of the Earth is left free from human interference. Elements of this sometimes sneak into the vision of sustainability—a world characterized by small, primarily rural and agricultural, self-sufficient communities; everyone living off-grid; minimal energy and material flows. These visions embed the same fallacy as Marx's: they fail to recognize the fundamental dynamics of complex systems. Absent a catastrophic collapse, this kind of much simpler future is highly unlikely. It is not a question of desirability, or even ethical preferability: many would rank Marx's utopia high on those bases. It is simply an observation based on history: it is highly likely that human society will continue to evolve toward greater complexity, not simplicity.

Even now we see the global governance system, through which environmental issues such as global climate change are being addressed, grow in complexity before our eyes: where once was only the dominant nation-state, now there are firms, NGOs, nation-states, and communities of all kinds. The dialog has gotten far more complicated, not simpler—and continued demands for transparency and participation by those still left out indicate that it will get yet more complicated.

That is not to say that individual elements of the system cannot become simpler, and certainly we can evolve ways of achieving quality of life without increasing consumption of material and energy. But it is to say that the system taken as a whole—the world as human artifact—will most likely continue to evolve toward greater complexity (indeed, modern cultural developments such as postmodernism, with its emphasis on cultural pastiche and the multiculturalism of modern political and social discourse, can be seen as a recognition of and response to the greatly increased complexity of human life just since World War II). The Marxist complexity fallacy is also a warning: the image of utopia, whether based on Marxism or sustainable development, is appealing in principle. But in practice it may not result in a Jeffersonian agrarianism, but in a Hobbesian state of nature: beastly, brutish, and short. Attractive ideas—community, simplicity—can become ideologies, potentially highly dangerous simplifications of the real world. Easy answers are usually wrong: if we are truly to do good, we must at least have the courage to understand the challenges we face—in all their complexity.

Changing of the Guard
(February 2003)

Recently, I got a pamphlet explaining how firms should embed social consciousness "into the DNA of their enterprises." Interesting. I have worked in a firm for a long time and have yet to see any DNA floating around the corridors (or, more relevantly, the corporate intranet). It is a strange turn of phrase, yet it sounds somehow comfortable and proper. What's going on?

From a cultural anthropology viewpoint, one of the most interesting dimensions of any society is the fundamental metaphor upon which its self-image is based. The metaphor is powerful, often simple, pervasive, obvious in some ways yet habitually unconscious. Thus, the reigning metaphor since the Industrial Revolution, from art through the conceptualization of the human organism, through the visualization of cities, has been the machine. Capitalist production, following the Taylor (human factors) efficiency and the Fordist mass production models, was viewed as a mechanism that could be tuned for efficiency—to the point where even Stalin and Mao, though they were following the siren song of Marxism, idolized Fordist production techniques. Where diplomacy broke down, "war machines" took over. And machine values—efficiency being the principle one—were paramount. And so forth.

But the past decades have seen a subtle but powerful shift of metaphor. We are now in the age of "ecotopia," of the "DNA of enterprises," of "eco-parks" replacing industrial estates. A powerful example is the language used regarding cities: the technocratic, machine language of Le Corbusier and Robert Moses has given way to the "ecological urbanism" of people like Abel Wolman, Gilles Deleuze, Felix Guattari, and Alberto Abriani. A highly contingent and idealized construct of "nature" becomes the blueprint for the city and the source of rules and underlying cultural validity for social structure: the landscape architect Ian McHarg, for example, asserts that the science of ecology applied to urban design provides "not only an explanation, but also a command."[1] With reference to business and the environment, the metaphor proves more than fertile: "natural capitalism," "industrial ecology," and "industrial metabolism." Make no mistake: one of the most powerful yet unrecognized sources of authority for environmentalism is its alignment with the metaphor of the age.

---

[1] I. McHarg, quoted in M. Gandy, *Concrete and Clay* Cambridge, MIT Press, 2002), 10. Gandy's book is an excellent discussion of the interplay of culture, capitalism, and "nature" in New York City.

Such metaphors are not inherently inappropriate. Just as the machine metaphor before it, the ecological metaphor is a useful way to view the world, and it provokes researchable problems of considerable interest. To what extent, for example, are industrial systems like ecosystems—and where do they differ? Industrial ecology has proven useful in beginning to model and understand larger-scale industrial flows of material and energy, and the field has grown to the point where it supports a competent technical journal and an international society.

The problems arise when the metaphor is mistaken for correspondence; when it is assumed that industrial or other human systems are, indeed, ecosystems in drag. And that is because the map is not the thing mapped: human systems are inherently of a different and higher order of complexity than natural systems. A salt marsh does not exhibit the contingent, self-referential behavior of New York City, where, for example, an observation about the city in the *New York Times* immediately becomes fodder for discussion, politics, and further analysis—and thus changes the initial observation. Nor, for that matter, is a large corporation a rain forest. Unquestionably, if a biological analogy is consciously used as a learning or educational device, it can be very useful—and it certainly speaks to the spirit of the age. The belief that biological systems are somehow equivalent to cultural and social systems, however, profoundly misunderstands the different degrees of complexity of natural versus human systems and can lead to seriously dysfunctional results.

Most importantly, perhaps, such a misunderstanding inverts the relationship between the cultural and the natural and in doing so makes rational management of significant environmental perturbations more difficult. "The environment" and, indeed, "nature" are cultural constructs, ideas that not only embed specific values, but reflect a particular time, place, and culture and are thus contingent and changing over time. To the extent they are regarded as absolutes, as somehow higher guides for behavior and social organization, they create the potential for significant cultural conflict and become paths for the introduction of rigid and powerful ideologies into environmental debates, with the usual consequences of unnecessary conflict and lack of real progress in addressing major issues. Fundamental metaphors are not bad: they have their uses, and are probably unavoidable anyway. But those who use them must always be very careful to remember the simple axiom: the map is not the thing mapped.

Resilience in Complex Systems
(January 2004)

As "sustainability" proves difficult to define precisely, some have increasingly turned to the concept of "resilience" as an alternative. While this tactic has the advantage of substituting a more concrete characteristic of many systems—robustness in the face of challenge—for the ambiguous term "sustainability," it is not enough by itself to provide analytical rigor.

The concepts of "resilience" and "vulnerability" are closely linked. Both refer to the sensitivity of complex systems to catastrophic failure from a minor change in input. Some complex systems are robust in that they do not tend to catastrophically fail with minor changes; others are vulnerable—as, for example, a surge in a single power line causing a massive blackout across the northern United States and parts of Canada. The same system may be fairly robust to a series of individual challenges, but eventually be pushed into a regime where a single additional change results in catastrophic failure. Thus, many biological communities are robust to initial challenge, but eventually a small additional challenge results in a significant shift in community structure and function (the proverbial straw breaks the camel's back). Obviously, this sort of behavior has serious implications for managing earth systems across all scales, from the local salt marsh that one wishes to keep healthy to trying to understand whether, and under what circumstances, global climate change might discontinuously shift oceanic circulation patterns, with potentially significant economic and demographic impacts.

Unfortunately, currently complex systems and their behavior are generally not well understood. In particular, how one should interact with complex regional and global systems of all kinds—from natural resource regimes such as the Everglades or the Baltic to human systems such as economies and technological systems—in order to maintain their stability in the desired ranges, is not at all clear. But research on complex systems and resilience at places like the Santa Fe Institute continues to deepen our understanding (readers might want to take a look at some of the relevant literature, such as Stuart Kauffman's *At Home in the Universe*; see this chapter's annotated bibliography).

Nonetheless, the focus on the resilience of complex systems can sharpen the nebulous sustainability dialog significantly. For one, it drives analysis toward the specifics of the systems in question, including their characteristics and behavior, and thus encourages intellectual rigor rather than rhetoric. Moreover, it challenges the static visions that sustainability can encourage, substituting the concept of self-organizing systems at the edge of chaos, characterized by energetically open and evolving systems continually building increasing levels of order and complexity. Resilience comes from growth and dynamics, producing a system that is constantly consuming energy to build itself and create new structures and behaviors and to maintain complex patterns of order, rather than energetically closed systems.

But shifting conceptually from sustainability to resilience in self-organizing systems is not enough. For the real question remains: resilience of what to what? More precisely, what characteristics of the system do we want to be resilient, and against what should they be resilient? This question should be considered in light of another important characteristic of complex systems: their relevant boundaries are defined by the query we pose.

Consider in this light the Internet. The resilience of such a scale-free network—which is indeed energetically open and self-organizing—to recover from random loss of links is extraordinary. On the other hand, a scale-free network is very susceptible to deliberate attack, because it usually has a limited number of hubs that are highly connected and form the backbone of the system. In short, such a network is very resistant to accident, and very vulnerable to attack. How we evaluate its resilience, therefore, depends on why we ask the question: are we interested in avoiding attack, or do we simply fear accidents? It requires us to dialog with the system in question, fostering a rigor of analysis and depth of knowledge that is otherwise all too frequently lacking from the discourse on sustainability. (The interested reader can find more detail in Albert-Laszlo Barabasi, *Linked*, described in this chapter's annotated bibliography).

And thus it is with many of the highly integrated human/natural systems, such as the climate system, biological systems at all scales, material cycles, and human systems such as cities. We need to understand their resiliences and vulnerabilities in detail, with explicit reference to the properties and behaviors we are particularly interested in. Beginning to ask hard questions about resilience and vulnerability is a step in a difficult but necessary direction.

Annotated Bibliography

1. D. H. Hofstadter, *Godel, Escher, Bach: An Eternal Golden Braid* (New York: Vintage Books, 1980). A well-written and creative introduction to many of the themes of complexity, done so gently that one hardly notices it. It remains one of the most charming ways to enter this confusing domain, retaining the sense of play that, in their seriousness, too many authors on the subject of complexity forget. Read it just for fun.

2. A. Barabasi, *Linked: The New Science of Networks* (Cambridge, MA: Perseus Publishing, 2002). One of the more interesting new areas of research is networks and network theory, and Barabasi's book is an excellent introduction (he is himself one of the principal researchers on complex network systems). The book is not mathematical and is clearly and concisely written so that the main points (such as why scale-free networks are resilient to accident but not to deliberate attack) are easily grasped.

3. J. Gleick, *Chaos: Making a New Science* (New York: Penguin, 1987) and M. M. Waldrop, *Complexity* (New York: Simon and Schuster, 1992). These two books are popular science treatments of their respective subjects and thus have the strengths and weaknesses of their genre. They are easy to understand, if somewhat chatty, and provide good introductions to their topics, albeit with the usual preoccupation with personalities that occasionally afflicts such works.

4. S. Kauffman, *At Home in the Universe* (Oxford: Oxford University Press, 1995). Kauffman is one of the pillars of the Santa Fe Institute, perhaps the major center dedicated to the study of complex systems, and a major thinker on self-organizing systems, especially in biology. This book is nonetheless quite accessible and easy to read, with no real demand for prior math and biological knowledge above that which a good college education in virtually any field should have provided. It is thus an authoritative and valuable introduction to different ways of thinking about complex systems and their dynamics.

5. J. M. Epstein and R. Axtell, *Growing Artificial Societies: Social Science from the Bottom Up* (Washington, DC: Brookings Institution Press, 1996). Epstein and Axtell are leading researchers in the area of agent-based computer modeling, where populations of agents evolve and interact based on small sets of simple rules. The behaviors that arise from these simple rules are often very complex and in interesting ways mimic trade, cultural expansion, and other types of human patterns. That such complexity can arise from simple origins is itself not necessarily intuitive, and it is useful to have these agent-based models illustrate the principle so convincingly.

Not only is this an interesting area of research in itself, but the agent-based modeling approach is being increasingly used in a number of fields, including industrial ecology, and so is a good methodology with which to be familiar.

CHAPTER 8

How Humans Construct Their Environment

As the previous chapter suggests, the overwhelming reality of the anthropogenic world is its complexity. Nonhuman systems are complex enough, but as the Anthropocene continues to unfold, human activities create new systems with emergent behaviors that are both unpredicted and, perhaps, unpredictable, as the ever-increasing integration of human and natural systems leads to a concomitant increase in importance of the particular forms of complexity that characterize human evolution. Thus, for example, biotechnology not only integrates genomic information of increasing numbers of species into human information systems, but into human economic and technological systems—not to mention political and ideological systems—as well. Biological systems thus become part of, and begin to exhibit, characteristics of human economic and cultural systems. This is not a new phenomenon: biological systems of many kinds have long reflected, directly or indirectly, human agricultural practices and technologies.

This points to a fundamental characteristic of the anthropogenic Earth: ethical, cultural, and religious systems are not just important abstractions to be studied by philosophers and sociologists. They are also, and perhaps primarily, critical forces shaping the physical world. The Everglades in Florida, the idyllic scenes of the Caribbean islands, the hinterlands of any major urban system, the northern European pastoral landscape characterized by small, "quaint" farms—these reflect the history of the dominant human cultures that created them. They are all constructed landscapes that look the way they do because of the human belief systems that have shaped these landscapes through history.

Similarly, the composition and structure of many island biological communities today at least indirectly reflects the culture of the European Enlightenment and the colonial ideology of eighteenth- and nineteenth-century Europe, which supported the evolution of a global transportation and trade system— and thus brought about extinctions of some species (such as the dodo) and spread many others (such as tobacco and sugarcane). Indeed, current biological communities are in many parts of the world as much cultural phenomena as products of "natural" dynamics.

This process is not limited to landscapes, although it is perhaps most visible in that context. Genetic engineering, a technology that will significantly restructure biological systems at all scales, has its roots in the historical European integration of Christianity and science and in the vision of technology as salvation. Bacon, for example, expressed this cultural value centuries ago in *New Atlantis*:

> We make (by art) in the same orchards and gardens, trees and flowers to come earlier or later than their seasons, and to come up and bear more speedily than by their natural course they do. We make them also, by art, much greater than their nature, and their fruit greater and sweeter, and by differing taste, smell, colour, and figure from their nature. . . . We find means to make commixtures of divers kinds, which have produced many new kinds, and them not barren, as the general opinion is . . . neither do we this by chance, but we know beforehand, of what matter and commixture, what kind of those creatures will arise.[1]

Bacon was not foreseeing genetic engineering as we know it. But he was certainly part of a culture that, more than any other, coupled religion and technology, became powerful thereby, and evolved a globalized culture—the Eurocentric culture that now dominates the world. And this culture, through genetic engineering, is now reshaping genomes just as it reshaped island biologies centuries ago.

In short, in the Anthropocene, human culture is reified over time in global systems. This does not occur directly, for these systems are too complex for such simple outcomes, but indirectly. As Gianbattista Vico wrote of human history:

> It is true that men have themselves made this world of nations, although not in full cognizance of the outcomes of their activi-

---

[1] F. Bacon, *New Atlantis* (MT: Kessinger Publishing Co.), 323–24.

ties, for this world without doubt has issued from a mind often diverse, at times quite contrary, and always superior to the particular ends that men had proposed to themselves. . . . That which did all this was mind, for men did it with intelligence; it was not fate, for they did it by choice; not chance, for the results of their always so acting are perpetually the same.[2]

To make things yet more interesting, the cultural dimension of the anthropogenic Earth also operates at different scales. Thus, culture and values become reified at a global scale as human activities change the evolutionary path of fundamental natural systems such as the carbon, nitrogen, and hydrologic cycles or the climate and atmospheric systems. This has happened over the course of millennia unintentionally and for the most part unnoticed, although that is changing: the climate change negotiations are, among other things, simply the latest mechanism by which cultural values are imposed on the climate system, with significant and for the most part unrecognized ethical implications (see "Ethics and Governance in a Postmodern World").

As the example of Bacon suggests, the process by which natural systems become expressions of human belief systems occurs over centuries of human development; it is not just a modern phenomenon. Although the scale and depth of this dynamic are certainly accelerating with increasing human population, advancing technology, and growing economic development, recent scholarship clearly demonstrates that even at the hunter-gatherer stage, human societies had substantial impacts on biodiversity, driving many prey and competitive megafauna extinct and substantially altering local estuarine ecologies. Recent scholarship suggests that as much as half of overall anthropogenic global climate change predates the Industrial Revolution. This was the result primarily of waves of deforestation and methane production from agriculture in Europe, northern Africa, and Asia. Belief systems matter, then, for they are the reason the world looks as it does—they are the reason it is an anthropogenic planet, and in many ways they bound the evolutionary paths for many forms of life on earth.

This is likely to be even more true looking forward, to the confluence of biotechnology with the other three foundational technologies: information and communication technology (ICT), nanotechnology, and cognitive sciences. Most people are blissfully unaware of the extraordinary scientific and technological evolution currently occurring: responsible researchers are predicting, for example, that within a few decades we will be doing such things as

---

[2] G. Vico, quoted in E. P. Thompson, *The Poverty of Theory* (London: Merlin, 1978), 291.

building cells from simple molecular constituents; controlling fighter aircraft remotely using wireless connections into our brain; living in a world with computational capability meshed across all geographic space, combined with comprehensive sensor systems and direct feeds into our cognitive structures; and rebuilding human bodies from the ground up (indeed, some foresee functional immortality within fifty years). Hidden in many of these scenarios is a redefinition of what it means to be "human," or what the boundaries of "human" are, which in turn suggests that the categories by which we currently define many systems will undoubtedly change profoundly in the future.

As with all scenarios, some may hold and others may not come to pass. But the implications for the structure of the human Earth are profound, and, partially because they reflect unfamiliar views of the world, not well explored. Consider as an example biodiversity, which virtually all environmentalists and ecologists see as diminishing significantly in the modern world. An alternative view, however, might suggest that rather than diminishing, biodiversity is in fact growing—only it is an anthropogenic biodiversity, generated from genetic engineering rather than the traditional biological evolution of species. And if that is the case—and there are many concerns about such a scenario—then biological systems will increasingly become a matter of our deliberate choice about what kinds of life we want (one might say "what kinds of species we want" except for the nagging thought that genetic engineering might render obsolete the concept of species, at least as regards designed life-forms).

It is here that the relative inability of the environmental discourse to recognize how value-laden it is could be unfortunate. At some point, failing to appreciate that environmental belief systems are highly culturally and historically contingent blinds us to the obvious fact that demography, culture, and technology refuse to stand still. This tendency assumes that the particular values of environmentalism should be dominant; it carries within it the potential to stifle discussion of alternatives and to grant insufficient respect to values that others may have (see "Values, Geography, and Environment," and "Environmentalism: Private or Public Morality"). Politically, such a tendency may remove an important cautionary voice from economic and technology policy dialogs.

As a specific discourse, it is perhaps appropriate that environmentalism reflect a particular ethical view of the world—most discourses tend to do so. But as environmental initiatives, such as the global climate change debate, expand to include a number of different discourses and values in addition to the purely environmental (such as jobs and employment, libertarian versus egalitarian approaches, acceptance of technological innovation, human health), the implicit ethical framework of environmentalism becomes inadequate. This does not necessarily reflect only on environmentalism, however, for there are as yet no ethical systems adequate to guide behavior and decision making in the Anthropocene. Indeed, the challenge of developing such a planetary ethical system goes far beyond environmentalism and points to a general failure of ethics, philosophy, and religion to generate a moral system that reflects the current, much less the future, state of the planet. While it is possible to suggest more-comprehensive ethical precepts in response to this dilemma (see "Toward a Planetary Ethic" and "Toward a Planetary Ethic: Part 2"), constructing a comprehensive ethics for the anthropogenic Earth is a work not yet in progress.[3]

Another difficult area that this understanding of the anthropogenic Earth highlights is that of "free will." This is a difficult enough concept where simple systems with understood cause-and-effect relationships prevail. It becomes far more complicated where complex adaptive systems are involved and where the results of choices may be neither predictable nor readily apparent. But if we understand that the world as it currently exists is increasingly a product of human activity and choice, we cannot avoid the question of how to ethically choose. This, also, is an area where some preliminary thoughts can be proposed (see "Free Will and the Anthropogenic Earth: One," " . . . : Two," and " . . . : Three"), work that has barely begun.

---

[3] These two columns on a planetary ethic are more high level than most others in this volume and do not do justice to the complexity of the issues involved. Further detail can be found in B. R. Allenby, "Observations on the Philosophic Implications of Earth Systems Engineering and Management" (Batten Institute working paper, Darden Graduate School of Business, University of Virginia, Charlottesville, 2002), discussed at more length in chapter 9's annotated bibliography.

Ethics and Governance in a Postmodern World
(April 2001)

The last few columns have discussed a number of subjects, including ethics, governance systems, Marxism and complexity, the Kyoto negotiation process, even cultural imperialism. At this point, some readers may be wondering how these topics tie together—and maybe even what they have to do with environmentalism. The answer, I think, is that taken together they begin to provide a clearer picture of a cultural phenomenon that, once necessary and beneficial, has become oversimplistic.

Postmodernists, whether in art, literature, architecture, or philosophy, are a strange lot. A little like medieval doctors, they have proven themselves excellent diagnosticians, but dangerous in treatment. Their diagnosis? A world grown so complex that it is understood by individuals in terms of ahistorical and ageographical pastiche: unlike cultures before us, we have access to and perceive bits and pieces of all times, all cultures, and weave them into something new. We live not in a place or a time, but in a process (think about how hectic your life is before you scoff too much at their diagnosis). The treatment? Governance based on the principle that no discourse is privileged—that is, that the cultural foundations that all of us rely on to some extent—religion, faith in science and technology, reason, Marxism—are only contingent, not absolute. Thus, for example, the writings of Michel Foucault and Jean-François Lyotard (and over all the shadow of Nietzsche), suggest that there are no absolute answers and that no set of beliefs can be imposed on others without (using Lyotard's term) a "terroristic" silencing of the other.

The treatment certainly goes too far, simply admitting defeat in the face of too much complexity, rather than attempting to come to grips with the messy reality of a truly multicultural world. There are dominant discourses in the world today—among them, technology and environmentalism as ideologies, not just policy structures. And yet the sensitivity of the postmodernists to cultural imperialism, to the silencing of those who may be different, is valuable both for ethical reasons and for pragmatic ones.

What can be more pragmatic, for example, than the inability of the global climate change talks to include the United States? While the blaming and finger pointing continue, and the specifics of negotiating issues are extremely complex, the foundational weakness of the Kyoto process is simple, and indicative of the failure of ideological environmentalists to understand the postmodern world within which they operate.[4] It is not that the science is not becoming clearer: virtually all the data, whether on patterns of coral bleaching or retreat of glaciers, tend to support the existence of some anthropogenic effect. Had the Kyoto process been just about responding to the science, it would have had a much greater potential for success. But the attempt to equate global climate change as a scientific phenomenon with the Kyoto process in a causal manner—the Kyoto process as the only rational reaction to the science as it now stands—is disingenuous. Rather, the Kyoto process is to a significant, albeit implicit, degree about the imposition of a specific worldview on a multitude of other voices and discourses: it is the exercise of a strong will to power by a disciplined and highly effective discourse. Such attempts often succeed—but not when the discourse that one is attempting to dominate is, in its turn, highly powerful (in this case, using a global climate change negotiating mechanism to attack the social structure, political values, and free-market ideologies of the United States).

The danger of this cultural warfare approach is obvious in the results: if the only vehicle that is encouraged is one embodying such conflict, and it is ineffective (as the odds are that Kyoto as now constituted will be), one is left with a policy vacuum. No viable alternatives have been created, for that is part of the implicit power play: my way or no way. Moreover, politicizing the dialog undermines the credibility of the associated science, which becomes seen as just an adjunct to one side of the conflict. But the irony of deep greens being allied with reactionary elements of the fossil fuel sector in destroying dialog is only superficial, for ideology on both sides gains when the difficult, complex multicultural dialogs of the common sense middle can be snuffed out. And in that, both the environment and the majority of the world's people lose.

---

[4] Many environmentalists and environmental groups are realists and are not as dominated by ideology as the deep greens tend to be. But it is the latter who, partially because of the power of their ideology, have been able to largely define the environmentalist discourse. This is in itself neither good nor bad: it depends on how one feels about the resulting governance system and its dynamics.

Values, Geography, and Environment
(January 2000)

Perhaps one of the most famous bumper stickers of environmentalism is "think globally, act locally." Often cited, often flouted, it is still a useful touchstone for thinking about modern environmentalism and environmental issues.

Begin with "think globally." Leave aside the obvious observation that it is necessary to consider earth systems at the appropriate scale—global for atmospheric dynamics and climate change; watershed for appropriate ecosystem and hydrological assessments; local for a toxic spill. Instead, concentrate on the political implications of that phrase. These are more problematic: for most parties to environmental debates, this translates into "think about how to impose my values on the world as a whole."

Capitalist institutions carry with them the ideology of free markets (and to some extent democracy and equality of opportunity, if not the egalitarian ideal of equality of outcome), and, more subtly, the Judeo-Christian Western worldview embedded in the discourse of modern technology. That environmentalism, particularly as practiced and institutionalized in the West, carries with it equally profound cultural values—for example, the Edenic mythos—is less recognized. In fact, one reading of the contretemps in Seattle around the 1999 World Trade Organization (WTO) meeting is that it represented a battle over how to develop trade levers that could be used to impose Western environmental values on other countries, particularly in the Third World. Indeed, some of the more technical discussions regarding imports of beef or GMOs, or protection of agriculture in the European Union, have equivalent ethical dimensions and raise powerful issues of values for the stakeholders involved. In short, the WTO meeting—unintentionally, one assumes—implicitly raised issues of fundamental values, of whose world we will live in. Such clashes are inevitable because virtually everyone thinks globally only in terms of their ideological and cultural preconceptions, their fundamental values. A clash happened in Seattle because there is no good place for such discussions on value to happen in the first place—and most people do not understand the extent to which their implicit values underlie their public positions.

"Act locally" is no less tricky. Two implications are particularly interesting. First, to some extent "act locally" is a challenge to the state and its technological and scientific apparatus. Particularly in the guise of high modernism, engineering and scientific enterprises and approaches powered by hubris and linked to state power have done significant environmental and social damage (consider the degradation of the Aral Sea in Asia, Soviet agricultural collectivization, or compulsory villagization in Tanzania). This model is still dominant in many cases, and to act locally, at the community level, challenges not just individual activities, but the culture behind the mask. This is particularly useful when community constraints and cultural patterns are most appropriate for preservation and use of localized resources.

But "act locally" carries with it dangers as well. The concepts of community and place embedded in it are superficially appealing and preferable to the cold, globalized, mechanized world, which is often the counterimage. But some of those technology systems have been the most powerful sources of individual freedom yet developed—think of the mobility and consequent freedom for women that access to automobiles provided, or the individual empowerment that contraceptive technology creates, or the intellectual and individual freedom that Internet access generates. These technological sources of freedom are powerful precisely because they are global. In interesting juxtaposition, the concepts of community and place can be dangerous, used as tools to impose authoritarian rule and ideological conformity and to discriminate, sometimes viciously, against "foreigners." In this regard, it is worth remembering that some of the fascist governments in Europe were among the most active in using "community" as a defining boundary of membership—and, indeed, were among the most environmentally active until many decades later. It is not that the idea of community is flawed; just that like any other human institution it can be misused. The dark side of communitarianism is inadequately appreciated by many of its adherents.

What is the common thread, then, between the global and the local—a thread that has wound its way through more than one of these columns? It is the power of value systems that are, for the most part, implicit and unconscious and thus all the more powerful. These value systems can be good, leading us to create possible futures of hope and plenty; more often, however, they generate conflict, providing superficial answers to complex questions and facilitating the demonization of any opposition. In the past, we have had the luxury of accommodating such conflict; now, the increasing complexity of the world and environmental perturbations is demanding a new maturity of all of us. We may as well begin with the Socratic admonition: "Know thyself." And thy values, one must now add.

Environmentalism: Private or Public Morality?
(November 2000)

Democratic systems of governance, especially those that guarantee true free-dom of speech and religion, impose a substantial responsibility on citizens. As individuals, virtually all of us have our own belief systems, reflecting our heritages, our social and cultural environments, our religious traditions, and, hopefully, our reflections on what is good and true. But a democratic system requires that we limit our efforts to impose those beliefs on others. Thus, in those societies where freedom of religion is a fundamental prin-ciple, such as France or the United States, believers in any particular faith are limited by the legal structure in the extent to which they can force others to follow their precepts. Such societies respect private moral beliefs but also require a more tolerant, public morality. A fundamentalist may practice his or her private morality, which may be quite rigid, but is expected to respect other belief systems that coexist in that society—a public morality of toler-ance and shared ethics. Those societies that do not reflect this difference between private and public morality—Afghanistan under the Taliban, for example—can be very unpleasant for those who do not hew exactly to the prevalent belief system.

Very interesting perhaps—but what does this have to do with environ-mental policy? Actually, a lot. Consider that most of the major environmen-tal issues we must deal with today—loss of biodiversity; global climate change; degradation of water, air, and soil resources—are global, if not phys-ically, then because they are integral parts of global economic systems. And the world, of course, is profoundly multicultural, as any transnational firm doing business in numerous countries can attest. There are indeed elements of global culture tending toward homogeneity, but these are interwoven with increasingly assertive local and regional cultural traditions, new ageo-graphic communities of interest arising because of the Internet, and the increasingly varied institutional cultures of firms and NGOs. A global gov-ernance system with any hope of managing this increasingly complex world must be, at least in terms of communication and participation, somewhat open, somewhat transparent—in other words, it will resemble in many ways a confused global democracy. Analogous to a democratic country, therefore, it will make the same demands for a differentiation between private and public morality, albeit in a new context: to what extent can, or should, an NGO, a firm, a society, or a country with strong "private" environmental moral beliefs impose them on other societies and cultures?

This conundrum is complicated by the attitudes and institutions generated by the adversarial history of environmentalism. Environmentalism in its early days needed to be ideological and powerfully emotional to overcome the barriers of existing practices and assumptions. Much environmentalism has succeeded to date precisely because private and public morality were conflated: protecting the environment was not just objectively good, but was an almost religious campaign against evil. More fundamentally, scholars have long noted the extent to which modern Western environmentalism is culturally constructed, reflecting elements of Christianity (the return to a primordial, sparsely inhabited, and definitely not urban Eden) and, in the United States, the frontier experience. These two unconscious but culturally powerful drivers contributed a great deal to the force of early environmentalism.

But the issues now are more complex and, to a far greater extent, lie beyond the boundaries of unitary cultures and single countries. Environmentalism has moved from an "overhead" issue—one to be dealt with only as an afterthought—to a "strategic" one, to be addressed as an important part of core concerns. In doing so, however, it is called to a new maturity. The very characteristics that gave early environmentalism its power and effectiveness in a national context now threaten to turn a progressive and effective movement into a largely unconscious agent of cultural imperialism and Western domination. Ideological rigidity becomes not a source of strength, but a powerful barrier to successful solutions—an enemy of the environment, as those who are excluded and damaged by such approaches find their voices and react in a backlash. The environment has indeed become an eternal issue, one that humans will have to be concerned about for the rest of their time on Earth. And yet, we will not be similar to each other; that is not our nature. Working out the language and nuances of an appropriate public and private environmental morality is therefore an important step in our acceding to the responsibility our activities have laid upon us.

Toward a Planetary Ethic
(February 2001)

In a previous column ("Environmentalism: Private or Public Morality?") , the differences between private and public morality as it applied to environmentalism were discussed. Private morality is that which we each have as differentiated individuals, reflecting our unique backgrounds, heritages, family situations, and the like, while public morality is the tolerance and shared ethics that each society requires of us in order to function. One classic type of social dysfunction occurs when there is a failure to appreciate the difference between these two. For example, when fundamentalists of any stripe take over a society, the results in today's information intensive, multicultural global system are both deplorable and unstable.

With the exception of some radical environmentalist groups—such as the Earth Liberation Front that has destroyed a number of buildings across the United States under the slogan "If You Build It, We Will Burn It,"—environmentalism taken as a whole reflects public, not private, morality. Public support for environmental protection—at least the forms that are most familiar such as local air, water, and waste pollution control—remains high. Moreover, polls indicate that a desire for a clean and healthy environment is not limited to developed countries, but is widespread across many cultures.

This is not the end of the story, however, for there is another level of ethics to which far less attention has been paid: planetary ethics. The need for this broader ethical base arises because, in their own way, public morality systems are also culturally and ideologically limited. Thus, as a citizen of the United States I have not only my individual ethical system, but also the shared ethical system—the public morality—of the United States. Further, in a profoundly multicultural world, members of other cultures and citizens of other states will have their own public moralities, and they will be different from mine. Analogous to the private/public morality dichotomy, it is both dysfunctional (in the sense of creating conflict) and arguably unethical (in the sense of cultural imperialism) for me to impose my public morality on someone else. (Obviously, as in cases of genocide or ethnic cleansing, there are circumstances justifying such imposition—but an appreciation of the ethical considerations raised by such an imposition is an important element of conducting such intervention in a defensible and constructive manner.)

In addition, there are ethical systems that have been created by certain communities, such as the environmental community, that are similarly limited in that they reflect bounded interests and concerns. Thus, for an environmentalist the prime ethical constraint may be "do nothing that harms the functioning of ecological systems"—and, indeed, many environmentalists may believe that this formulation (or something similar) is the basis of all ethical systems. But this is a public morality, not a planetary morality, for it is grounded in only one community reflecting only one set of interests (albeit an important set). For example, it says nothing about an entire class of issues involving interpersonal ethics: the way individuals treat each other where no environmental values are at play. Moreover, it gives no guidance in cases where environmental ethics may conflict with other values and a prioritization among such values is required: is it ethical for an individual to harm an ecosystem if it is the means by which (s)he feeds a family? It represents, in other words, a public but not a planetary ethic. And yet, particularly in light of the scale of human impacts on the environment, as captured in the concept of earth systems engineering and management (ESEM), there clearly is need for a fundamental ethical base that does not discriminate against groups or individuals based on their stage of development, discourse, religion, or culture. In short, ESEM and the globalization of environmental issues require a planetary ethic—but environmentalism to date has produced only a public ethic.

But, in light of the postmodernist critique that there are no privileged discourses, is trying to develop a planetary ethic simply feasible? It must be admitted that any ethic that mandates specific behavior probably cannot be considered as planetary. To the extent an ethical system reflects only one discourse (for example, market capitalism or environmentalism), culture (as in European positions on GMOs), or religion, it cannot be considered a planetary ethical system. But that does not mean there are not similarities across cultures, and that those cannot be used to formulate a planetary ethic. Indeed, that is what I shall try to do in the next column.

Toward a Planetary Ethic: Part 2
(March 2001)

The last column suggested the need for a planetary ethic; this column rushes in where angels fear to tread and suggests a possible formulation. This is not done in a vacuum; in fact, considerable work has been done, particularly in the Parliament of the World's Religions, which, based on the work of the theologian Hans Kung, developed a Declaration Toward a Global Ethic.[5] As part of that document, Kung notes the universality of the Golden Rule (citing on pages 71–72 formulations in Confucianism, Islam, Buddhism, and Hinduism, among others). Accordingly, and in recognition of the need for an ethical foundation for earth systems engineering and management (ESEM), I derive a possible planetary ethic from that base (in particular, the formulation of the Categorical Imperative as developed by Kant): "Act such that the world that would be expected to result if every entity acted in an equivalent manner would be an ethical and desirable expression of human design."[6]

Needless to say, there are important elements of this formulation that require some explanation. To begin with, "act" reflects the point that, in a world where important dynamics of natural systems are dominated by human activities, there is no such thing as inaction. To fail to act is to eschew responsibility, but not to avoid the impacts that humans are having now and will have in the future. "The world" reflects the fact that, for the anthropogenic Earth, natural and human systems are so intertwined that they cannot be separated. In the real world, environmental issues are intertwined with scientific, technological, cultural, social, institutional, and natural systems in such a way that the network of these relationships, not the specific issue, is the relevant analytical, and ethical, whole.

"Would be expected to result" reflects two different issues. First, the internal dynamics of complex systems lead them to evolve in ways that are unpredictable, sometimes even in theory, when perturbed. Second is the question of how much is actually known about the system; this is the slightly different issue of knowledge, data availability, and concomitant uncertainty regarding the future behavior of the system. I may in theory be able to predict how the system will evolve, but still be unable to predict evolution in

---

[5] See H. Kung and K. Kuschel, *A Global Ethic* (New York: Continuum, 1998).
[6] I. Kant, *Grounding for the Metaphysics of Morals*. Translated by J. W. Ellington, Cambridge: Hackett Publishing Company, 1993, 30.

fact because of a lack of knowledge or data. Under such circumstances, actors cannot be required to predict with absolute accuracy the systematic effects of their actions. It is, however, reasonable to demand that they consider what is most likely to result given the state of knowledge and the internal structure of the system.

Most ethical systems—and formulations of the Golden Rule—assume implicitly a universe populated only by individuals. The anthropogenic Earth, however, is characterized by a multitude of potential actors at very different scales: the individual, the tribe, the firm, the NGO, the state, the class, the world spirit (*geist* of Hegel), and (for many religions) God or deities. "Every entity" recognizes that it is insufficient to simply consider the appropriateness of individuals, as opposed to actors at all scales, aligning with one's ethical decisions. The deeper—and currently unanswerable—question of whether such actors are "conscious" need not be addressed under this formulation.

In complex systems, one cannot simply wish other actors to behave the same, for they may have different roles, functions, constraints—or, for that matter, cultural values. "An equivalent manner" recognizes this complexity; it is the underlying fabric of the planetary ethic that must knit together different kinds of activities into an ethical whole.

"Ethical" is taken in the sense of not violating the particular ethical standards that, not yet completely formulated, are necessary to inform the evolution of the anthropogenic world. "Desirable," on the other hand, represents the need for procedural equity, the process of nondominated human discourse. Cultural imperialism is occasionally viewed by its practitioners as ethical—remember the European colonialists' "white man's burden." But it is not, under the planetary ethic standard, "desirable," for it requires the stifling of dissenting dialogs and discourses.

Finally, we refer back to "human design." The rationale for this exercise is, after all, the need to evolve existing public ethics, including environmental formulations, into the planetary ethic necessary for the anthropogenic world. And central to that project is the realization that the world is, in ways perhaps too wonderful and complicated to understand at this point, increasingly a product of human design.

Free Will and the Anthropogenic Earth: One
(October 2004)

Jalalu'ddin Rumi, the twelfth-century Persian poet, made the pertinent observation that "there is a disputation [that will continue] till mankind are raised from the dead between the Necessitarians and the partisans of Free Will."[7] The observant reader will note that the specified event has not occurred, and, indeed, there remains no closure on whether free will exists or what it is. A brief column on environmental issues is unlikely to resolve this issue—so why on earth bother?

Very simply, free will matters in large part because in most cultures—including the dominant Eurocentric globalized culture—ethical responsibility accompanies decisions made where free will exists and does not accompany actions that do not arise from free will. Thus, moral responsibility is generally not imposed where actions are taken under duress or by individuals who are incapable for some reason, such as mental illness, of exercising free will.

Extending this principle into the complex, self-organizing structure of an anthropogenic Earth is not trivial. It is undoubtedly true that even given an Earth dominated by the activities of one species, many ethical issues continue to arise in the realm of individual decisions, where traditional formulations (and arguments) about free will prevail. But such a world also exhibits emergent behaviors at high levels of the system, and there the prevailing models break down for a number of reasons, many of which arise from the fundamental unpredictability of emergent behavior and related questions of "control" (if one controls a system, one may be held responsible for its behavior; if one does not or cannot, imposition of ethical responsibility is problematic).

Consider, for example, the Portuguese ship engineer who built the first caravel (a light sailing ship of the fourteenth century that was particularly adept at sailing into the wind and accordingly was the basic design that enabled European colonization). If the initial boat had sunk because of bad design, few would have qualms about assigning moral responsibility to the engineer. However, it strikes most people as inappropriate to assign responsibility for the subsequent colonization process, with all of its social and environmental dimensions, to the individual engineer: there are simply too

---

[7] Rumi, quoted in R. Kane, *The Significance of Free Will*, Oxford: Oxford University Press, 1998, 3.

many intervening decisions and stochastic events. Or, to take a more recent example, consider the environmental and social implications of the Internet. It is a complex system that is clearly entirely human in origin, for every piece of it—from routers, to transmission infrastructure, to personal computers used to access it—is of human design and manufacture. On the other hand, the Net itself has been designed by no single individual or institution; indeed, there are not even any good maps of the Net, for it continually redesigns itself. It is a self-organizing system. The Net as a whole may have "good" or "bad" features, but these effects are far beyond the designer of a particular component of the Net.

Microethical systems—ethics at the level of the individual as a member of a particular culture or profession—are not free of disagreement and complexity, but at least it is well-tilled ground. Macroethics, however—the ethics of emergent behavior of technology systems, societies, or intercultural relations that a profoundly multidisciplinary world calls forth—is an area that to date has yet to receive adequate attention. Such consideration requires not just new ethical formulations, but new institutional roles—if the individual engineer is not ethically responsible for the emergent behavior of the Net, then who is?

These questions become particularly important in light of the rapid evolution of new foundational technologies, particularly those known as NBIC—nanotechnology, biotechnology, information and communications technology, and cognitive sciences. These technologies almost certainly will transform human beings, their cultures, and the anthropogenic Earth and its systems: researchers talk of functional immortality within fifty years; of building cells from simple molecules in two decades; of seamless wireless human/remote machine systems within several decades. Specific scenarios may or may not occur, but our current evolutionary trajectory carries with it enormous ethical questions and implications. While I will suggest several ideas in the next column about how we may begin considering some of these, a far more basic message is clear: these futures are not hypothetical, but are in the process of being reified; we are not prepared for them; we have little time to begin doing so; and our current ideologies are ill-suited to these unprecedented and complex challenges. Luxuriating in past verities is no longer just irrelevant, but is itself unethical.

Free Will and the Anthropogenic Earth: Two
(November 2004)

My previous column discussed the relationship of free will to ethical responsibility, noting a weakness in the area of macroethics, the ethical structures applicable to large complex economic, social, environmental, and technological systems. This weakness becomes important given the integration of human and "natural" systems that characterize the anthropogenic, or "human-generated," Earth, especially at larger scales (consider, for example, the Everglades; the climate system; technological systems such as nanotechnology, biotechnology, information and communications technology, and cognitive sciences; and urban systems and their sometimes global hinterlands). How ethical systems for such challenges—with issues that cut across very different systems of values (such as employment versus environment) and that involve strongly disparate distributions of benefits and costs—are to be formulated, much less managed and institutionalized, has yet to be effectively addressed.

Unfortunately, this is not just a matter of "scaling up" traditional dialogs about free will and ethics, which tend to focus on individuals. Historically, the two bases for judging the ethical posture of an action have been by the intentions behind, or by the actual consequences of, the action. But such traditional approaches implicitly assume a simple system structure, where enough can be known at the time a decision is taken to be able to impute moral responsibility for the results of that decision. If, instead, the systems are inherently unknowable, at least beyond a trivial point, then I can neither have an honest intention as to what I hope to achieve (because the complexity of the system response means my intention is essentially irrelevant, since whatever I want is unlikely to occur), nor can I be judged by the consequences, which are beyond my ability to determine and become apparent only over significant time frames.

This has several important implications. Although the locus of free will remains the individual, the exercise of free will becomes a function of the state of the system within which the individual is located. Free will becomes a question of context, not just an inherent characteristic of a human being. As an important corollary, this means that the interconnectivity and internal dynamics of complex systems become additional constraints on the exercise of individual free will. Complex interlinking technology systems limit option spaces within which free will can be exercised.

These characteristics of free will in complex systems lead directly to another implication: ethical implications adhere less to specific choices and more to the choice of mechanism by which we choose to interact with the relevant system. Macroethics thus differs from microethics in requiring a greater concern with processes, as opposed to single actions. Micro- and macroethics each has its role, but it is a fatally flawed category mistake not to recognize their differences.

For example, if I am designing the Everglades—and, in doing so, trying to balance human development, agricultural and mining interests, and environmental interests such as avian biodiversity—the results of many individual decisions are difficult to determine except as the response of the overall system becomes clear. Therefore, whether judged by my intention or by the consequences of any particular choice, any individual action is itself meaningful in an ethical sense only as it becomes reified in the system with which I am interacting. Thus, my choice of the process by which I engage in dialog with the system itself is what becomes ethically critical. For example, I may choose any one of a number of particular actions—channelizing a stream, planting a marsh to reduce phosphorous concentrations in agricultural runoff—but because the potential outcomes of each action become clear only as the system—including its human components—adjusts, I am behaving unethically if I do not monitor the results of my action and change my actions accordingly. In other words, if I choose not to adopt a process that fits the system, I have behaved unethically, for in doing so I have deliberately undermined my ability to exercise free will, and thus my ethical nature.

Free will and ethical responsibility in complex systems such as the Everglades thus become less single-point functions and more like networked functions spread over multiple spatial and temporal scales. Just as quantum mechanics did not render obsolete Newtonian physics, but relegated it to a limited space (for example, interaction of macrobodies), the traditional concept of free will is thus not obsolete, but is a bounded part of a much more complex, systems-based phenomenon. My next column will suggest some practical implications of this analysis.

Free Will and the Anthropogenic Earth: Three
(December 2004)

The previous columns have argued that traditional conceptualizations of free will and accompanying ethical responsibility must be expanded in two important ways. First, they must comprehend not just individual moral judgments, but they must be able to guide uncertain and highly complex actions affecting complicated integrated human/natural systems. Second, they must begin to focus less on the ethical implications of individual actions resulting in predictable outcomes and more on process, on continued dialogs with complex systems with even the desired outcomes (the earth systems engineering and management [ESEM] design objectives and constraints) contingent and changing over time, as the ideologies and political positions of stakeholders and affected entities shift. This leads to a final question: how do we get from where we are to where we want to be, a macroethical system capable of guiding actions under such unprecedented conditions?

We can begin by rejecting the common approach of simply projecting individual ethical responsibility to the scale of emergent behaviors of ESEM systems. It is simply untenable to make individual scientists or engineers responsible for the behavior of systems to which they may have contributed, but which in many cases are self-organizing and demonstrate behaviors that are unpredictable and that become apparent only over significant time periods. The designer of a chip that goes into a router that goes into the Internet cannot be held personally ethically responsible for how the Net may affect social structures thirty years from now; nor can she, using new 65-nanometer chip technology, be held responsible for the social effects of nanotechnology. The complete inability to predict what such effects may be; the multitude of events and decisions between the chip design and eventual social and cultural responses; the tenuous connection between specific individual design decisions and overall system responses . . . all operate to reduce the causal linkages between individual choice and outcome upon which ethical responsibility rests.

Pushed to its limit, such an ethical posture, similar to the strong construction of the precautionary principle, is simply a mechanism to attempt to freeze technological evolution—and, indeed, that position is advocated by some based on such an analysis. The pragmatic response is that history provides few examples where technologies have been successfully halted (as opposed to modified, regulated, or controlled, which has happened frequently); where it has occurred, another culture has almost always stepped in to develop and exploit the technology. Moreover, taking the extremist view that technology should be stopped by imposing on individuals the ethical duty not to participate, while ideologically satisfying, tends to remove an important source of potential constructive criticism from the social dialog.

But the individual scientist, engineer, or environmentalist can, it seems to me, be charged with a fundamental responsibility to establish a process by which technical communities, and society at large, can engage in dialog with complex technological systems and by which ESEM communities can do the same with earth systems such as the Everglades, the climate cycle, or New York City. The nature of such a dialog, which must be highly multidisciplinary and multicultural, is itself a reason why individuals cannot carry such a burden in a substantive sense, for no single individual has the requisite knowledge, and very few have the ability to suspend their own ontologies, as such a dialog requires. Individuals of all kinds—from engineers, to scientists, to environmentalists—can, however, certainly be charged with ethical responsibility for supporting the procedural process. The dialog itself will have to rest with an institutional host—one that combines technical knowledge with a broad, transparent, and open process and that is sensitive to its own agendas and ontologies and can be explicit about them without imposing them on the dialog. Moreover, such dialogs should be, and should be seen to be, relatively safe from capture by a particular religious or political agenda, a problem that some have noted with regard to stem cell research in the United States, for example.

Thus, we would charge our engineer with the ethical responsibility to push her professional organizations—such as the Institute of Electrical and Electronic Engineers (IEEE), the National Academy of Engineering (NAE), the AAAS, or perhaps in some circumstances an academic institution or a national laboratory—to create an institutional framework within which an ongoing macroethical capability is established. In so doing, we begin to move toward a framework that remains based on individual free will and ethical responsibility, but that reflects the increasing complexity of the problems, options, and constraints that characterize the anthropogenic Earth.

Annotated Bibliography

1. P. Anderson, *The Origins of Postmodernity* (London: Verso, 1998). Post-modernism is a difficult discourse to define; it can be thought of as aes-thetic Marxism, as opposed to the usual economic variety, and emphasizes the themes of cultural pastiche and ethical (indeed, ontological) rela-tivism. Regardless of what one thinks of postmodernism, it is important to understand something about it, for elements of the postmodernist dis-course have become important in many relevant areas, from discussions about sustainability to questions of how and whether it is appropriate to impose Western values on other cultures. In this regard, those who have a high tolerance for Marxist rhetoric might want to explore M. Hardt and A. Negri, *Empire* (Cambridge: Harvard University Press, 2000), which as part of its postmodern analysis of modern power structures identifies European and American human rights and environmental NGOs as the "frontline force of imperial intervention" of the Euro-American empire. Most Americans and Europeans will be surprised and perhaps bemused by this claim; it is important to recognize that there are many in the devel-oping world who find the idea quite realistic.

2. I. Hacking, *The Social Construction of What?* (Cambridge: Harvard Uni-versity Press, 1999). This is an excellent introduction to the whole vexed question of construction (and deconstruction) of culturally contingent concepts and belief systems. It is valuable not just as an aid in understand-ing the contingent nature of many environmental beliefs, but also in explaining in fairly simple and logical ways the conflict between scientists and technologists, and their postmodernist critics.

3. M. H. Abrams, *Natural Supernaturalism: Tradition and Revolution in Romantic Literature* (New York: W. W. Norton and Co., 1971). In many ways, the current environmental discourse is a direct descendant of Rousseau's idiosyncratic interpretation of Enlightenment Romanticism. More generally, that element of environmentalism that believes that non-human nature constitutes the sacred reflects the transmutation of Chris-tianity in the Enlightenment as science and rational thought began to replace authority as the foundational source of knowledge. That the envi-ronmentalist discourse should thus reflect Christian utopianism should be no surprise given its roots. But most students of the environment are not students of the European Enlightenment, and vice versa, which explains why an excellent book on Romantic literature finds its way into this bibliography.

4. M. Berman, *All That is Solid Melts into Air* (New York: Simon and Schus-ter, 1982). In this interesting book (with a title drawn from a famous pas-

sage from Marx), Berman explores the cultural landscape of modernity and postmodernity in an easily accessible way with a minimum of jargon. Whether and when modernity fades into postmodernity is less important than understanding the fundamental differences between the high-technology state-sponsored projects typical of the former, as compared to the more open, community-oriented politics typical of the latter. A particularly good commentary on high modernity and its costs, by the way, can be found in J. C. Scott, *Seeing Like a State: How Certain Schemes to Improve the Human Condition Have Failed* (New Haven: Yale University Press, 1998), which is also highly recommended.

5. H. Kung and H. Schmidt, eds., *A Global Ethic and Global Responsibilities: Two Declarations* (London: SCM Press, 1993). This volume, edited by a prominent Catholic theologian and a former chancellor of Germany, presents two draft documents attempting to establish an ethical framework for a complex, multicultural world. One is entitled "The Declaration of the Parliament of the World's Religions," and the other "A Universal Declaration of Human Responsibilities." It is an ambitious attempt, and anyone concerned with the difficult question of ethics and the environmental and sustainability discourses should be aware of it. Indeed, my construction of a planetary ethic draws on the effort, but expands it to reflect a focus on systems complexity, thereby moving it somewhat away from a fairly anthropocentric structure to include, for example, bioethical approaches.

CHAPTER 9

Implementing Earth Systems Engineering and Management

The previous chapters have revolved around the fundamental reality of the Anthropocene and the profound challenges that this new situation poses to both environmentalism and human society generally, especially when we are in the early stages of trying to adapt to it. But laying this out can have a daunting effect that enables the theoretical best to become the enemy of the pragmatic good. It is important to realize that we can begin by doing a lot better than we are doing now, even while we grapple with the more difficult social, scientific, philosophical, and ethical dimensions of a rapidly changing and increasingly anthropogenic planet. What is needed is serious engagement at all temporal scales, from the relatively rapid time cycles of policy development and deployment to the much longer ones of foundational social and cultural change.

Accordingly, this last chapter lays out some initial earth systems engineering and management (ESEM) principles and themes that can guide better practice in the short term, drawing on and integrating lessons from several existing areas of study. These include learnings from not only well-established disciplines and fields such as law, theology, philosophy, sociology, and economics, but also multidisciplinary fields such as industrial ecology, adaptive management, urban studies, and systems engineering. Moreover, there is a large and growing literature on developing and deploying complex technology systems—such as the Internet, global financial and transportation systems, military weapons systems, and the like—and the reflexive relationships of such systems to society, that provides useful experience—and cautions. Based on such sources, one can suggest some preliminary ESEM principles that can be sorted into three general categories: theory, governance, and design and management.

But first let us begin with a cautionary example. A classic example of unintended impacts (and of high-modernist hubris, for that matter) is the case of the Aral Sea, which is actually a lake straddling the borders of Kazakhstan and Uzbekistan. Only decades ago it was the fourth largest lake in the world, but in a few short years it lost about half its area and some three-fourths of its volume because more than ninety percent of the flow of two of its feeder rivers, the Amu Darya and the Syr Darya, was diverted to irrigate arid land to grow cotton for the Stalinist Soviet Union. The resulting irrigation system is extremely inefficient. Some estimates are that only 30 percent of the diverted water, carried in unlined canals through sandy desert soils, reaches its destination. The unintended results of this engineering project are profound and staggering. Much of the region has shifted from arid to desert; the resulting Ak-kum Desert is expected to reach three million hectares by 2005, although it did not even exist 35 years ago. Some 40 to 150 million tons of toxic dust per year with substantial detrimental impacts on regional agriculture and human health are now generated every year, with potential impacts on the climate regimes of China, India, and southeastern Europe. Reduced river flow has led to increased salinization of the Aral Sea and a concomitant loss of twenty of its twenty-four fish species and a drop in the fish catch from forty-four thousand tons in the 1950s to zero today (meaning the loss of sixty thousand jobs). Avian species have been decimated, with a reduction in nesting bird species in the area from 173 to 38. On top of everything else, there is a continuing concern that biological warfare agents previously contained because they were quarantined on an island (Voskreseniye Island) may be released as the Aral Sea shrinks, creating instead a peninsula.

The lesson here is important: hubris, especially when combined with the power of a state (and especially when that state lacks internal checks and balances), is a significant danger to both responsible ESEM and to the systems that may be part of such efforts. It is simple realism to recognize that technological evolution and economic activity are going to continue in the Anthropocene; but it is quite another to succumb to unquestioning technological optimism. A good ESEM practitioner will value her or his sense of ignorance and humility—and be alert to potential ideological prejudices and the challenges of a particular historical period (see "King Cotton on Biomass" and "Biomass Management Systems" for some thoughts on how one might approach what is perhaps the oldest ESEM practice—agriculture). This observation is also pertinent, as some environmentalists appear to equate ESEM with a completely artificial world and then proceed to attack the concept on that basis: it is important to realize that such a straw man is not at all what ESEM is about.

Having recognized this danger, let us turn to an initial set of ESEM principles.
Theoretical Principles: The theoretical underpinnings of ESEM reflect the complexity of the systems involved.

1. Because integrated human/natural systems are complex, uncertain and unpredictable, intervene only when necessary, and then only to the extent required. Minimal intervention reduces the probability and potential scale of unanticipated and undesirable system responses. Note, however, that this is only one guiding principle: its extension into a commandment, as some purport to do with the precautionary principle ("don't introduce any new practice or technology until and unless you can show it has no risks or costs associated with it") is both unrealistic and oversimplistic in a world of six billion people and continuing cultural, economic, and technological evolution.

2. ESEM projects and programs are highly scientific and technical in nature—but they also have powerful economic, political, cultural, ethical, and religious dimensions as well. All of these facets should be explicitly integrated into ESEM approaches. In particular, the tendency to rephrase issues of values in strictly technical or scientific terms, and to disregard that which cannot be quantified, should be resisted.

3. As a corollary, ESEM projects often combine technical scientific and engineering issues and efforts to change behavior (social engineering). This is not necessarily inappropriate, but every effort should be made to differentiate between the two: the discourses, political contexts, and degrees of complexity involved are quite different.

4. ESEM requires a systems-based approach, with analysis and boundaries reflecting real-world behavior and characteristics rather than disciplinary or ideological oversimplicity. Thus, while specific design and management activities may be part of an ESEM activity, they are not equivalent to ESEM design and management. For example, structuring the Everglades in light of continuing human pressures will no doubt involve significant civil and mechanical engineering design of the hydrologic systems—but these are not substitutes for the ESEM design and management required at the overall regional system level.

5. The way problems are stated defines the systems involved. Accordingly, ideology will often be implicit in the way problems are defined, rather than explicit. For example, an environmentalist might define the Everglades system by asking "How can we return the Everglades to its original biological diversity?", while a developer might ask "How can we manage water supplies so as to avoid negatively affecting built communities and ensuring adequate water for future growth?" By thus establishing an

implicit ideological approach, each has bounded the system that would be involved in responding to the question in different, and oversimplistic, ways. The ESEM approach requires that, prior to asking such questions, the system itself—including relevant stakeholders—be understood as a whole. In this way, boundaries subsequently drawn around it should reflect real-world couplings and linkages through time, rather than disciplinary or ideological simplicity.

Governance Principles: One must begin this section by being honest: in many cases, how exactly these principles can be implemented, especially given current practices, is not apparent. But these principles establish goals toward which both individual policies and policy development and deployment practices can move, and at least the last one offers an immediate opportunity to do much better than we currently are.

6. The complexity of integrated human/natural systems, the broad scope of most ESEM initiatives, and the danger of cultural imperialism create a need for consensus and transparency, which can be met only by governance processes that are open, democratic, transparent, and accountable. For these reasons, ESEM activities at all stages should be characterized by an inclusive dialog among all stakeholders rather than the traditional control-oriented approach. How this may be done in practice is an open question: examples such as the United Nations are as cautionary as they are encouraging. But that it should be done, for both ethical and pragmatic reasons, is clear.

7. ESEM governance models must be flexible and able to respond quickly and effectively to changes in a system's state and dynamics; this will require including the policy maker as part of an evolving ESEM system, rather than as an agent outside the system guiding or defining it. This raises the difficult policy issue of matching the cycle time of the systems that are the subject of ESEM, and the policy and governance mechanisms that purportedly are intended to address them. In many cases today, for example, it is clear that the government entities that should be addressing technical, environmental, social, and ethical implications of new technologies (or cultural developments platformed on them) operate on cycle times that are far longer than the underlying phenomenon, making policy prescriptions essentially irrelevant.

8. Especially as human adjustment to an anthropogenic world is just beginning, it is particularly important to ensure that continual learning at the personal and institutional level is built into ESEM processes. Thus, society

should not only be working on ESEM projects such as atmospheric design and global climate change management and Everglades design and operation, but should also be deliberately learning from the experience. This sounds relatively easy, but in practice is extremely difficult: national and international research and development structures are by and large predicated on disciplinary boundaries, and funding, peer review, professional status, and the like are all tied to disciplines. Multidisciplinary work, the essence of learning in the Anthropocene, is very much the exception rather than the rule.

9. Because ESEM projects usually continue over time (it is difficult to foresee when humans will not have to be working with the climate cycle, or hydrologic cycle, or biological systems at all levels, absent catastrophic population collapse), it is important to ensure that adequate resources, over time, are available for support of both the project and the associated science and technology research and development. Particularly with complex adaptive systems, this long-term commitment is necessary to ensure that the responses of the relevant systems are understood as they occur.

10. Major shifts in technologies and technological systems should, to the extent possible, be explored before, rather than after, implementation of policies and initiatives designed to encourage them. For example, encouraging reliance on biomass plantations as a global climate change mitigation effort or new energy source should not be implemented at scale until predictable implications on critical systems—nitrogen, phosphorus, and the hydrologic cycle, for example—are explored. Such exercises should not only look at technical and environmental implications, but also the ethical questions that might arise. And in all cases such exercises should become part of a dialog with the evolving system, rather than just one-off studies. If properly organized, academic institutions, professional organizations, or national laboratories might be positioned to undertake such work.

Design and Engineering: The assumption that humanity is capable of rational and ethical design and engineering of earth systems remains an assumption—albeit the alternative assumption, that humanity stumble along until something happens, seems rather defeatist. Indeed, there are some principles in this area that can be added to the more general ones above.

11. ESEM is a story of emergent characteristics at high levels of system organization. Accordingly, evaluations of technology should also include evaluations of scale, and scale-up should allow for the inevitable (especially in

complex systems) discontinuities and emergent characteristics. This is especially important given the growth in human population and consumption, the globalization of the economy, and the scale of commodity markets. A technology that works well at small scale—say, the use of arsenicals to protect wood from insects—may have quite different effects when commercialized in a huge globalized market.

12. When working in an ESEM process (say, managing the Everglades or the Baltic Sea), explicitly identify the overall design objectives and constraints ("maintain avian diversity in the Everglades at X levels"), and establish quantitative metrics by which progress can be tracked. Equally important, identify metrics and markers by which potential negative systems behaviors can be identified and addressed as early as possible. With ESEM projects, the process is understood as an ongoing dialog, rather than an effort to simply achieve a designated endpoint. Responsibility for the system never ceases.

13. Policy, design and engineering initiatives in ESEM systems should be incremental and reversible, rather than fundamental and irreversible; "lock-in" of inappropriate or untested design choices as systems evolve over time should be avoided whenever possible. Where lock-in appears likely, its implications should, to the extent possible, be understood and properly managed beforehand.

14. ESEM engineering and management should attempt to foster resilience, not just redundancy, in relevant systems (redundant systems have backup mechanisms for a particular subsystem or functions, but are still subject to difficult to predict catastrophic failure, while a resilient system resists degradation generally and degrades gracefully when it must). As a corollary, inherently safe systems (designed to fail in noncatastrophic ways when they do fail) are to be preferred to engineered safe systems (designed to reduce the risk of catastrophic failure although there is still a finite probability that such a failure may occur).

These principles are generic and incomplete, and will certainly be augmented and improved as we go forward, for, although the anthropogenic world is not a new phenomenon, our perception of it is relatively recent and incomplete. Although the challenges of such a planet are daunting, we must remember that despite the torrid pace of change there is no reason to believe that this necessarily means an ending of anything. Rather, it is the beginning of the next stage of our evolution: to be in continued dialog with the creation for which our species is responsible, and part of, and indeed to a large extent has already begun to define. As we begin this process, it is important to recognize from the beginning that it is as much cultural, and human, as it is scientific and techno-logical or, in fact, "environmental." In this light, ESEM is not the beginning of the anthropogenic Earth, but the rational and ethical response to that which our species has already begun to create.

King Cotton on Biomass
(February 2004)

A recent article in *The Economist* provides an interesting CV for the crop known as King Cotton: highlights include a primary role in beginning the Industrial Revolution (circa 1750–1820); forming the economic basis for slavery and the Civil War in the United States (circa 1840–1865); destroying the Aral Sea and much of the surrounding areas (circa the 1960s to the present); and fueling the failure of international trade negotiations in Cancun (2003).[1] An impressive bio indeed—but it also leads to more general observations in an era when biomass generally is being touted as an "environmentally friendly" solution to all sorts of problems, from carbon sequestration, to energy production, to a nonfossil sources of plastics and other materials. The next two columns will explore some of these.

Let us begin with the obvious. As the Aral Sea environmental disaster illustrates—much of the flow of the two major rivers feeding that body of water was diverted by the Stalinist Soviet Union to grow cotton—agriculture has for millennia been perhaps the major human activity through which human and natural systems became integrated over time. Indeed, the need to establish and manage large hydrologic engineering systems to provide water for agriculture has been identified by historians as a major factor in the rise of many civilizations (China is a prime example of such a hydrologic civilization). And global climate change through increases in anthropogenic carbon dioxide in the atmosphere began with the deforestation of Europe, north Africa, and parts of Asia a millennium ago as agriculture spread, not just with modern technology (although the magnitude of that change has obviously increased recently). Conversely, for some environmentalists such as Dave Foreman of Earth First!, the "fall" of the human species occurred precisely at that point where human society evolved from hunter-gather "natural" patterns to "unnatural" agricultural systems. If there is one activity that can be identified as a crucial factor in the evolution of our planet into what *Nature* calls the anthropocene, it is agriculture.

The current tendency to idolize biomass as an environmentally friendly feedstock for energy and material production is thus somewhat contradictory to begin with. It of course reflects naïveté about the systemic effects of agricultural activity; many programs encouraging enhanced production of biomass fail to consider effects on coupled systems such as local and regional

---

[1] See "A Great Yarn," *The Economist,* December 20, 2003, 43–46.

hydrologic and carbon cycles, much less the more esoteric and less familiar nitrogen and phosphorous cycles. Such mythic thinking also, however, reflects something deeper. Agriculture is associated with food production, and few materials are as surrounded with taboos, rules, cultural practices, and ideology as food. Moreover, agricultural activity, and concomitant cultural patterns, are a critical connection of many modern peoples to a (largely imaginary) past. The Americans, the British, the French, the Germans: all to some extent romanticize an idyllic time of small farms and agrarian yeoman, despite actual agricultural practices in developed countries having become highly industrialized. This is no passing fancy: the Romans (and the Europeans that interpreted them) valued the farmer-citizen of the Republic over the decadent urban dweller of the Empire, and the modern quasiutopianism of many sustainability activists mirrors such an agrarian teleology. It is not just raw economics that keeps subsidies so high in the agricultural sector and inflicts such inequities on developing countries trying to compete in global biomass markets—it is a powerful, often implicit, cultural yearning for the simplicity of the agricultural golden age past. Mythical, to be sure, but still a powerful vision. The lesson from King Cotton is not just "fear me, I am economically powerful," but "fear me, for I am culturally potent as well."

Thus agriculture, like other great technological systems, functions not just as a production mechanism in a global economy. Rather, agriculture illustrates the complexity of such systems in the real world, where natural and human systems are integrated at all scales. It also illustrates the interplay of powerful and unconscious cultural archetypes with short term policies and political posturing, as cultural constructs such as "nature" and "environment" are created and renewed through the lens of technological systems and social values (for example, organic farming as "natural" and therefore "good"). But this still begs what to many is the critical question: what form of agriculture is "best," or "sustainable"?

Biomass Management Systems
(March 2004)

As last month's column noted, agriculture is the most substantial and certainly the longest-running human experiment in earth systems engineering and management (ESEM). It is thus not surprising that it is in the agricultural discourse that foundational questions, such as the relationship between human design and nonhuman life-forms, are first arising. The direct effects of agriculture are virtually definitional: modification of both floral and faunal species to better provide human benefits, be they food, fiber, fuel, clothing, or the like. The indirect effects on biodiversity have been, if anything, more significant given the massive changes in human demographics and settlement patterns, infrastructure systems, consumption habits, and economic and technological systems enabled by agriculture over human history. That it should be the agricultural sector where biodiversity begins to become directly mediated by humans in the form of GMOs—with all the attendant cultural conflict—is thus, from a historical perspective, predictable.

But the current debate over GMOs and agriculture is in some sense a look in the rearview mirror, reflecting the application of increasingly anachronistic worldviews to a future that promises to be significantly different than the present. Consider, for example, the fundamental definitional question of the role of agriculture in the future. No doubt, it will continue to provide everything it does now—but it is not unlikely that demand for biomass of all types, produced using agricultural technologies, will expand significantly. This has several important implications. First, questions of integrating environmental, social, and economic efficiency in agricultural operations and technologies will increase in importance. Second, agricultural activities will increasingly become critical leverage points where human and fundamental natural cycles—hydrologic and climate systems; carbon, nitrogen, and phosphorous cycles; land-use patterns; human cultural, economic, and technological systems—integrate. Accordingly, agriculture will in part shift from being a "cause" of environmental perturbations to being a management technique for regional and global natural systems. Thirdly, this critical integrative role will reflexively change both technology and the culture. Thus, substantially more reliance on GMO technology may become necessary as land use, production efficiency, and adaptation to changes in climate systems occur. More subtly, critical cultural constructs may also shift. For example, "biodiversity" might evolve from being a reflection of patterns in an external environment ("What species are out there?"), to being a reflection of human design ("What modified life forms have we designed?").

This is not to say that such changes will be "good" or "bad," only that current trends indicate that they may be probable. For we currently lack solid data and analytical frameworks to make assertions about the costs and benefits, and resulting normative assessments, of different kinds of agricultural practices. Is developed-country "organic" farming good or bad compared to the alternatives? Objectively, the data are too sparse to support an answer. Thus, for example, organic produce may not have some pesticide residues, but it may have a higher fungal load, with concomitant higher concentrations of potent fungal toxins such as aflatoxin (produced by *Aspergillus flavus*). Similarly, no-till agricultural practices can have environmental benefits (for example, less erosion), but these increase (significantly?) if GMOs, which facilitate weed management, are used. Organic production technologies may be appropriate for some specialty crops, but might require much more land if used for core crops, such as corn. Economically, organic food is a luxury item, which carries with it certain equity considerations (if organic food is healthier overall, is it to be only for the rich?). And such questions beg entirely the implications of agricultural needs and practices in developing countries, which might be entirely different.

Ideological biases frequently substitute for analysis where data and comprehensive assessment methodologies do not exist. What is needed, however, is not better bumper stickers, but rigorous and systemic evaluations of biomass production processes, taking into consideration both existing and potential future demand and supply and technological options, and looking at environmental, social, economic, and cultural costs and benefits comprehensively. Such studies should not try to reach normative solutions—whether organic cotton is better or worse, for example—but should try to identify the relevant dimensions and factual considerations that society and individuals might consider in reaching their own conclusions. The point of such an ESEM approach to the question of appropriate biomass production processes is not to close on any single answer at this point, but rather to develop the information and wisdom critical for the successful continuing evolution of agriculture in the anthropogenic world.

Annotated Bibliography

1. There are no books I am aware of on earth systems engineering and management (ESEM); the most comprehensive discussion is in a working paper (and it is of uneven quality): B. R. Allenby, "Observations on the Philosophic Implications of Earth Systems Engineering and Management" (Batten Institute working paper, Darden Graduate School of Business, University of Virginia, Charlottesville, 2002). This is being expanded into a book, but remains the most comprehensive source on ESEM.

2. V. Smil, *Cycles of Life: Civilization and the Biosphere* (New York: Scientific American Library, 1997). Smil is a prolific and easy to read writer, and this book is a good overview of natural systems and their relation to human systems. It is neither highly technical, nor so watered down that it fails to interest and educate, so it is a good resource.

3. "Managing Planet Earth," special issue, *Scientific American* 261, no. 3 (1989) and "Human-Dominated Ecosystems," special report, *Science* 277 (1997): 485–525. Both of these issues, as one would expect from these journals, are excellent. The *Scientific American* covers a broad range of issues, from forestry to climate, and remains a valuable read, although it is obviously somewhat dated. It has the distinction, however, of containing the article by Robert Frosch and Nicholas Gallopoulos that many believe was instrumental in beginning the study of industrial ecology. The *Science* special report is more technical, and more focused on biological systems, but its message—that "there are no places left on earth that don't fall under humanity's shadow"—is perhaps all the more powerful for that.

Index